Benchmarking and Self-Assessment for Parliaments

DIRECTIONS IN DEVELOPMENT
Public Sector Governance

Benchmarking and Self-Assessment for Parliaments

Mitchell O'Brien, Rick Stapenhurst, and Lisa von Trapp, Editors

WORLD BANK GROUP

ISBN (paper): 978-1-4648-0327-7
ISBN (electronic): 978-1-4648-0328-4
DOI: 10.1596/978-1-4648-0327-7

Cover image: Created via Wordle.net.
Cover design: Naylor Design.

Library of Congress Cataloging-in-Publication data has been requested.

Contents

Chapter 6   **Benchmarking for Democratic Parliaments**   **127**
                  Anthony Staddon and Dick Toornstra

Chapter 7   **Parliamentary Benchmarks: A Requisite for Effective**
                  Official Development Assistance   **139**
                  Alice French

Chapter 8   **The Role of Parliamentary Monitoring Organizations**   **155**
                  Andrew G. Mandelbaum and Daniel R. Swislow

Boxes

Figures

Tables

Foreword

The Commonwealth Parliamentary Association (CPA) is proud of its role in the development and promotion of benchmarks for good parliamentary practice. The influence of the CPA's landmark publication of *Recommended Benchmarks for Democratic Legislatures* in 2006 has now spread far beyond the 53 countries of the Commonwealth, with many other parliamentary associations having subsequently drawn up their own sets of criteria, often drawing heavily on the pioneering work of the CPA.

The impetus for change came initially from without, rather than within. Towards the turn of the millennium, aid agencies and donors, in particular, started to recognise the value of good parliamentary governance. Potential programmes supporting parliaments had to be assessed in the same way as, for example, projects building hospitals. This led to the United Nations Development Programme (UNDP) publishing its own set of parliamentary indicators in 2001.

Meanwhile, members of parliaments and parliamentary officials were becoming increasingly alarmed that standards were being drafted by those who did not understand how legislatures actually worked in practice. Shopping lists of good governance ideas from donor agencies, academics, interest groups and the executive arms of governments could not be allowed to dominate the parliamentary reform agenda.

The turning point came at a 2005 meeting on parliamentary benchmarking in Washington, DC, convened by the CPA and the World Bank, and attended by donors as well as intergovernmental and parliamentary groups. After Washington, parliamentarians took control of the agenda. The meeting participants agreed that the World Bank and the CPA would hold a study group of parliamentarians and parliamentary clerks to develop benchmarks against which parliamentary assistance could be measured.

The resulting benchmarking study group, hosted by the Parliament of Bermuda, with support from the UNDP and the European Parliament, drew upon previous CPA work and codified many previously unstated understandings into the 2006 benchmarks.

Recommended Benchmarks for Democratic Legislatures has subsequently proven to be a practical tool for the improvement of parliamentary processes across the Commonwealth. And there are excellent examples in this book of how the benchmarks have operated in an overwhelmingly positive manner. Beyond mere assessment, the benchmarking process has given parliamentarians an opportunity to reflect upon and improve the operations of their parliaments whilst widening interparliamentary discussions.

In response to this broadening debate on parliamentary standards, the CPA has continued to push the benchmarking agenda, first by encouraging its Member Parliaments to assess themselves against the *Recommended Benchmarks for Democratic Legislatures*, and then with the development of Regional Benchmarks for the Pacific; Caribbean, Americas, and the Atlantic; Asia; India; and Southeast Asia. In addition, a number of these regional benchmarks reflect the particular conditions affecting legislatures in those parts of the world. And it is this approach—because of the sheer diversity of legislatures in the CPA, which includes national and subnational parliaments, one of which represents more than a billion people and some with barely a thousand voters—that has been one of the CPA's greatest strengths in promoting parliamentary change.

Parliamentary standards will continue to evolve in response to pressures from both inside and outside of parliaments themselves. The CPA has already put its study group process to work to develop further benchmarks to enhance good governance and build trust in Parliament. Parliamentary openness, social media, and the behaviour of parliamentarians have all been identified as areas where the benchmarks need to be developed and refined.

Before the CPA benchmarks, parliaments and parliamentarians were used to being constantly assessed by outsiders. The benchmarks finally allowed parliamentarians to assess themselves against standards developed by their peers. Assessment and scrutiny of parliament will no doubt continue to increase over the coming decades. But through the hard work of its members and member parliaments, the CPA will continue to ensure that parliamentarians remain at the heart of the reform process.

Akbar Khan
Secretary-General
Commonwealth Parliamentary Association

Preface

This book comes at an opportune time, with international focus on improving governance institutions and country systems. Across the globe, grassroots movements show us that there is universal desire by citizens for greater transparency, participation, and government accountability in order to build trust between communities and the governments that serve them. Effective parliaments are a critical component in ensuring this accountability and transparency. Periodic elections that reflect the will of the people are no longer sufficient to hold the executive accountable and maintain the trust of citizens. With the renewed emphasis on strengthening country systems and improving transparency and accountability as part of the international partnership on aid effectiveness, parliaments have become an even more important player in the development equation. This is coupled with a growing number of parliaments globally that are seeking to assert their independence from the executive so that they are better able to perform their critical representative, legislative, and accountability functions. There has been a renewed interest from parliaments and members of parliament (MPs) to manage the increasing demands of better governance. It is increasingly important that legislatures must now play a vibrant role in ensuring that elected MPs respond to evolving citizen expectations.

The World Bank's Governance Global Practice (GGP) seeks to contribute to the Bank's twin objectives of ending extreme poverty and boosting shared prosperity by helping client countries to enhance governance systems and build more inclusive institutions. Enhanced transparency, participation, and citizen-engagement in decision making and stronger accountability contribute to more inclusive institutions that are trusted by citizens to deliver responsive and effective governance. Central to these efforts is our global and country engagement programming that seeks to strengthen parliamentary institutions. In order to ensure that the World Bank's interventions are evidence driven, an innovative approach that combines data collection and applied research with practitioner-focused initiatives aimed at surfacing and distilling global norms and practice, has been employed.

Assessing parliamentary effectiveness requires some form of criteria and measurement of performance. Over the past 15 years, the World Bank, along with its partners, developed effective benchmarking frameworks for parliamentary evaluations. These exercises have been found to be beneficial for parliaments to self-evaluate their performance. These frameworks also provide a useful tool by which development partners can have a conversation with parliaments as to how they can collaborate in order to strengthen parliamentary performance.

The publication of this edited volume represents, by no means, the end of the global discussion around what constitutes parliamentary good practice. It is intended, however, to provide a snapshot of how far we have come and form the basis for an ongoing dialogue around how parliaments and development partners can work together to enhance parliamentary performance, thereby strengthening participation, transparency, and oversight in country systems.

I would like to thank Finland's Ministry of Foreign Affairs, as part of the World Bank–Finnish Parliamentary Partnership, and the Commonwealth Parliamentary Association for their support of the World Bank's Parliamentary Strengthening Cluster and the publication of this volume.

Jeffery M. Thindwa
Practice Manager, Governance and
Inclusive Institutions
The World Bank

Acknowledgments

The publication of this volume marks the 10-year anniversary of the start of an international process aimed at building consensus around parliamentary benchmarks. The growing importance of this agenda is tied to the emerging body of evidence linking parliamentary performance with effective, accountable, and inclusive institutions and more peaceful and inclusive societies. This volume brings together the various initiatives and streams of work undertaken by several associations and organizations over the past decade. It aims to be a compendium for members of parliament and practitioners interested in the different options and approaches for benchmarking parliaments. Therefore, the contents of this book draw heavily on the extensive work led by numerous organizations and individuals who over the past decade saw value in refining how we support and measure the success of parliamentary-led reform initiatives.

This volume could not have been written without the tireless efforts of numerous parliamentary associations and regional parliaments who facilitated the direct participation of parliamentarians from different regions and parliamentary systems in several parallel processes. In particular, we would like to herald the leadership of the Commonwealth Parliamentary Association, the Inter-Parliamentary Union, the Parliamentary Assembly of La Francophonie, Parliamentary Confederation of the Americas, the European Parliament, and the Southern African Development Community Parliamentary Forum in ensuring that parliamentarians had a clear voice in the global discussions as to what characterizes a high-performing parliament and the best approaches for assessing parliamentary performance.

I would also like to acknowledge the institutional partners that we have been collaborating with over the past decade on different benchmarking initiatives. These include the United Nations Development Programme (UNDP), United States Agency for International Development (USAID), the United Kingdom Department for International Development (DFID), African Centre for Parliamentary Affairs (ACEPA), International IDEA, Parliamentary Centre of Canada, National Democratic Institute (NDI), State University of New York (SUNY), Global Organization of Parliamentarians Against Corruption (GOPAC), Westminster Foundation, Deutsche Gesellschaft für Internationale Zusammenarbeit (GIZ), the Organisation for Economic Co-operation and Development (OECD), and the Parliament of Finland.

We would also like to thank the many chapter contributors, many of whom are affiliated with the parliamentary associations, regional parliaments, and institutional partners that we have collaborated with over the past decade. These include David Beetham, M. Pascal Terrasse, Jaques Chagnon, Anthony Staddon, Dick Toornstra, Alice French, Andrew G. Mandelbaum, Daniel R. Swislow, Rasheed Draman, Raja Gomez, Jill Anne Joseph, Wayne Berry, Tom Duncan, Hon. Taomati Iuta, and Jennifer Smith.

Many colleagues at the World Bank and the CPA helped shepherd this initiative to completion. Without their dedication and motivation, this compendium would not have been possible. In particular, I want to thank Vineeth Atreyesh Vasudeva Murthy, Miriam Bensky, Niall Johnston, and Sruti Bandyopadhyay for their tireless efforts. I would also like to express my thanks to the CPA team members with whom we have collaborated so closely—in particular, Meenakshi Dhar, Lucy Pickles, and Arlene Bussettee.

We would like to extend our deep appreciation to the Finnish Ministry for Foreign Affairs, which, as part of the World Bank–Finnish Parliamentary Partnership, supported this initiative. I would also like to thank the Dutch Government through the Bank-Netherlands Partnership Program (BNPP) for providing earlier financial support for analytic work on parliamentary benchmarks and indicators.

Finally, I would like to thank my fellow editors, Rick Stapenhurst and Lisa von Trapp, both of whom were present when this decade-long journey began. They have been key players in stewarding this debate and nurturing a global process that has embraced diversity in practice between different regions and parliaments from different parliamentary traditions and recognized the richness that this diversity brings to the global discussion.

Mitchell O'Brien
Senior Governance Specialist, Governance Global Practice
The World Bank

Abbreviations

ACT	Australian Capital Territory
AG	auditor general
AGI	actionable governance indicator
AIDS	acquired immune deficiency syndrome
APF	Assemblée Parlementaire de la Francophonie (Parliamentary Assembly of La Francophonie)
API	African Parliamentary Index
APSP	African Parliamentary Strengthening Program
ASGP	Association of Secretaries General of Parliaments
CAA	Caribbean, Americas, and the Atlantic
CDI	Centre for Democratic Institutions
CoE	Council of Europe
COPA	Parliamentary Confederation of the Americas
COPE	Committee on Public Enterprises
CPA	Commonwealth Parliamentary Association
CPC	Commonwealth Parliamentary Conference
CSO	civil society organization
DFID	U.K. Department for International Development
FPOCC	Forum Presiding Officers and Clerks Conference
GAC	governance and anticorruption
HIV	human immunodeficiency virus
HRM	human resource management
ICA	independent country assessor
ICT	information and communication technologies
IDEA	(International) Institute for Democracy and Electoral Assistance
IMCIF	Initiative Multilatérale de Coopération Interparlementaire Francophone (Multilateral Initiative for Francophone Interparliamentary Cooperation)
IPO	interparliamentary organization
IPU	Inter-Parliamentary Union

LALT Network	Latin American Network for Legislative Transparency
LTTE	Liberation Tigers of Tamil Eelam
MAF	Management Accountability Framework
MP	member of parliament
NDI	National Democratic Institute for International Affairs
NGO	nongovernmental organization
ODA	official development assistance
OECD	Organisation for Economic Co-operation and Development
OGP	Open Government Partnership
OIF	Organisation Internationale de la Francophonie (International Organization of La Francophonie)
OSCE PA	Organization for Security and Co-operation in Europe Parliamentary Assembly
PAC	public accounts committee
PACE	Parliamentary Assembly of the Council of Europe
PBO	Parliamentary Budget Office (Canada)
PEFA	Public Expenditure and Financial Accountability (indicators)
PILDAT	Pakistan Institute of Legislative Development and Transparency
PMO	parliamentary monitoring organization
PSC	Public Service Commission
SADC PF	Southern African Development Community Parliamentary Forum
TI	Transparency International
UNDP	United Nations Development Programme
URL	uniform resource locator
USAID	U.S. Agency for International Development
XML	Extensible Markup Language

Introduction

Mitchell O'Brien

Summary

Over the past 15 years there has been an international effort to improve the effectiveness of parliaments. International organizations, such as the World Bank and the United Nations Development Programme (UNDP); bilateral aid agencies, such as the United States Agency for International Development and the United Kingdom's Department for International Development; and international parliamentary associations, such as the Commonwealth Parliamentary Association (CPA) and the Inter-Parliamentary Union (IPU) have come to realize the importance of parliaments for good governance.

At the same time, popular movements—such as the Arab Spring in the Middle East and the large anticorruption protests in India, which have pressured the Indian parliament to enact a bill to create an ombudsman office—reflect growing public aspirations for more accountable governments. In response, public institutions, including parliaments, are keen to be more open, responsive, independent, and accountable. In particular, many parliaments are adopting reform measures that lead to enhanced strategic planning and more modern approaches to corporate management.

These two separate developments—increasing focus on parliamentary strengthening within the international community and adoption of reform programs by many parliaments—are resulting in a common demand for parliamentary performance indicators and benchmarks. International development assistance organizations need to demonstrate to the governments that fund them (and, ultimately, to the taxpayers) that their legislative assistance programs are yielding positive results. At the same time, parliaments themselves need baseline indicators against which they can judge their own reforms. The result has been the emergence of a number of different, albeit complementary, approaches to assessing parliamentary performance.

The World Bank and the CPA hosted the first international meeting of scholars and practitioners interested in the development of parliamentary performance indicators in December 2004. Participants called for more substantive research, including a review of the literature on the subject, the results of which were considered by a study group that was supported by the CPA, the World Bank, the UNDP, and the National Democratic Institute for International Affairs (NDI). The outcome was the publication of the world's first benchmarks for democratic legislatures (CPA 2006). Since then, a plethora of additional approaches and variations on this issue have been published, including by NDI; l' Assemblée Parlementaire de la Francophonie (the Parliamentary Assembly of La Francophonie, or APF); the Southern African Development Community Parliamentary Forum (SADC PF); the Parliamentary Confederation of the Americas (COPA); the Parliamentary Centre; and the IPU. In addition, many regions of the Commonwealth, including Africa, Asia and the Pacific, and the Caribbean, have adapted the CPA's initial set of benchmarks to better reflect their own sociopolitical, cultural, and historical contexts. However, attempts to forge an international consensus (for example, at an international seminar and workshop organized by the World Bank and Griffith University that was held in Brisbane, Australia, in 2008) proved elusive. In particular, tension arose regarding the choice of tools that could be used: (a) standards or benchmarks against which parliaments could be compared and assessed; (b) self-assessment tools, which could help parliaments assess where they stood; and (c) performance indicators, which could help parliaments track their performance against planned objectives.

This book presents a *tour d'horizon* of the current state of parliamentary indicators and benchmarks. Though not trying to be prescriptive—there are already sufficient approaches to benchmarking—this book presents a comprehensive overview of the various approaches now in use around the world by both international organizations and parliaments. It also describes some of the public pressures that have arisen and are requiring parliaments to improve their performance. Moreover, the book presents some case studies of how parliaments are using the various assessment tools to improve their performance. It should be noted, however, that the world of parliamentary benchmarking and performance indicators is rapidly changing, with parliamentary reform occurring in many regions of the world.

Outline of the Book

Within the various benchmarks approaches, a high-level consensus exists regarding the core functions of the parliament: representation, lawmaking, and oversight. Members of parliament (MPs) *represent* their constituents in the parliament and create laws on behalf of citizens. Representation includes conducting public outreach and fulfilling constituency responsibilities. Benchmarks require that parliamentarians have the resources and facilities to represent their constituents effectively: they should have access to technology to effectively communicate with their constituents to enable citizen engagement in governance. Benchmarks also highlight that the media's access to the parliament should be transparent and nonpartisan.

Lawmaking ranges from ritualistic legislative involvement to complete partici-pation in governance processes as constitutionally mandated (Stapenhurst and others 2008). The evaluation criteria in the APF's framework propose that the parliament should have clearly established procedures to table bills and amend-ments, review them, and enact them. Furthermore, these procedures should regulate discussions such that MPs have opportunities to debate the bill or amendment before it is put to a vote. Parliamentary committees perform a sub-stantial portion of parliamentary work and scrutinize government expenditure, review policy decisions, and examine legislations. It is thus important that legisla-tion be referred to committees for detailed review and analysis (see chapter 4 by M. Pascal Terrasse).

Parliamentary *oversight* refers to "the review, monitoring, and supervision of government and public agencies, including the implementation of policy and legislation" (Yamamoto 2007). Benchmark frameworks agree that parliaments should have the right to oversee the decisions and actions of the government. Budgetary oversight is crucial to review government's budget and expenditures, and benchmarks require that a nonpartisan supreme audit institution table audit reports to the parliament in a timely manner.

Systems of Benchmarks

This book begins by comprehensively analyzing the work to date on developing assessment frameworks for parliaments. In chapter 1, Lisa von Trapp outlines four of the most commonly cited frameworks: the NDI's International Standards for Democratic Legislatures, the CPA's Recommended Benchmarks for Democratic Legislatures, the APF's *critères d'évaluation*, and the IPU's Self-Assessment Toolkit for Parliaments. Von Trapp highlights the commonalities and differences among the frameworks: there is a high degree of consensus in themes such as institutional independence, procedural fairness, transparency of parlia-mentary information to the public, core parliamentary functions and powers, and procedural fairness and internal democracy, but approaches to political financing, party discipline, and specific mechanisms historically associated with the type of parliamentary system differ.

Anthony Staddon and Dick Toornstra, in chapter 6, analyze the rationale behind parliamentary benchmarks and suggest how they can be operationalized. Analysis of country-specific contexts should include a review of the political background; constitutional and international rights and obligations; relationships among the parliament, the executive, and the judiciary; public perception and public access to parliament; and socioeconomic, cultural, and traditional contexts (IFES 2005, 7). For the benchmarking exercise to be useful, parliaments that have used the frameworks need to follow up on the recommendations and implement them. Political will and leadership have to exist within the parliament to improve its effectiveness. Parliaments may be reluctant to measure their per-formance because they fear that their "bad" practices will be exposed. Furthermore, parliaments do not generally quantify information such as changes of legislation, cost savings, or improvements in service (CCAF-FCVI 2004, 10).

The Evaluation Criteria

With this context of frameworks, in chapter 2, David Beetham reviews the uses of the IPU's Self-Assessment Toolkit, the issues covered in the self-assessment process, and cases of countries that have used the toolkit. The toolkit offers a framework to identify the main strengths and weaknesses of the user's parliament against widely accepted criteria for democratic parliaments. As Beetham demonstrates, parliamentarians can use the toolkit to formulate priorities for improvement and assess the effectiveness of reforms already in progress. He provides an interactive analysis of the toolkit and invites readers to self-evaluate their parliament.

Chapters 3 to 5 present a collection of progressive criteria and objectives in different regional parliamentary associations, such as the CPA, the APF, and COPA. In chapter 3, Akbar Khan, Secretary-General of the CPA, showcases the development of the CPA's benchmarks through the study group process, which enabled a group of experts to recommend good practices with respect to various aspects of parliamentary functioning. Omorodion reminds us of the differences in opinion among the CPA and partner organizations on regional benchmarks and their effect on governance. In chapter 4, M. Pascal Terrasse presents a collection of criteria and objectives for APF parliamentary members to strive for, notwithstanding the fact that the APF is a collection of diverse parliaments, and in chapter 5, Jacques Chagnon provides the recommended benchmarks for the COPA. Building on the efforts of the APF, the CPA and the IPU, Chagnon notes that COPA developed its own criteria, comprising four main sections: (a) elections and the status of parliamentarians, (b) parliamentary prerogatives, (c) the organization of parliament, and (d) parliamentary communications.

In chapter 9, Rasheed Draman introduces the African Parliamentary Index (API) and covers the API's purpose, scope, and methodology. This self-assessment exercise was undertaken in five countries in Sub-Saharan Africa (Benin, Ghana, Kenya, Tanzania, and Uganda) and covers five core areas (representation, legislation, oversight functions, institutional capacity, and institutional integrity). The questions, which are quantitative in nature, help users of the API to undertake comparative analyses of different countries and highlight good practices and lessons learned.

Trends in Governance and Benchmarks

Chapters 7 and 8 look at the larger trends in parliamentary benchmarking. The international aid landscape has moved from undertaking large projects to providing aid using a partnership approach, with responsibility for how funds are used assigned to the recipient partner. In chapter 7, Alice French makes a case that the international development assistance community needs to develop benchmarks that encompasses allocation, monitoring, and post-implementation review of government expenditure funded from development assistance. If legislatures are strengthened, she suggests, not only will aid dollars be used more efficiently, but also improved governance will ultimately reduce aid dependence and promote self-sufficiency.

Along with the trend in international aid and the emergence of benchmarks in parliamentary organizations, citizen-based groups have begun to focus on the role of parliaments, especially in the oversight function in democracy. At present, more than 220 parliamentary monitoring organizations (PMOs) monitor slightly more than 90 national parliaments worldwide. PMOs work toward collaborative governance by engaging citizens in the legislative process. In September 2012, the declaration on parliament openness was launched, taking into account nearly 130 organizations in 75 countries to call on parliaments to increase their commitment to openness and citizen engagement. Andrew G. Mandelbaum and Daniel R. Swislow, in chapter 8, articulate the roles and effects of PMOs, the importance of the Declaration on Parliamentary Openness, and the benefits of greater collaboration between PMOs and parliaments on normative frameworks.

The Cases of Sri Lanka, Canada, and Australia

Chapters 10, 11, and 12 analyze the use of different benchmarks in specific country contexts. In chapter 10, Raja Gomez examines Sri Lanka's parliament using both the IPU parliamentary indicators and the CPA benchmarks. He concludes that Sri Lanka's problems do not stem from a lack of experience in legislative procedures but from a resource shortfall. He notes significant outcomes, such as participants' recognition of the need for constitutional and electoral reform. If self-assessment exercises are viewed as a technique for identifying priorities and a means for strengthening parliament, parliamentarians will find them beneficial.

In 2009, the Parliament of Canada completed a self-assessment using the CPA benchmarks. Jill Anne Joseph analyzes the Canadian experience in chapter 11 and identifies key areas in which benchmarks should be developed: (a) governance and management; (b) parliamentary information and public outreach; and (c) legislative, oversight, and procedural functions. She proposes benchmarks in each of these areas in a parliamentary and accountability management framework and discusses how to devise an assessment scale.

In chapter 12, Wayne Berry, a former speaker of the Australian Capital Territory (ACT) legislature, and Tom Duncan, a clerk of the ACT legislature, share the main findings of the CPA benchmarking exercises they each undertook, respectively, in 2006 and in 2011. The ACT legislature not only enhanced its performance in those areas that were identified for improvement in 2006, but also found that both parliamentary staff members and elected members better understood the purpose of benchmarking for good governance.

Given the rapidly changing world of parliamentary benchmarking—and especially the adaptation of generic frameworks and models to fit particular regional and national perspectives—it is impossible either to present a completely up-to-date snapshot of parliamentary benchmarking around the world or to draw valid and meaningful cross-country comparisons. Instead, this book summarizes the evolution of parliamentary benchmarks and performance indicators over the past 15 years; compares and contrasts different frameworks and models (noting that the different approaches typically reflect different objectives and, thus, are not necessarily mutually exclusive); and presents some case studies of how some forward-looking parliaments have used the benchmarks. No single publication can aspire to present a completely up-to-date snapshot of parliamentary benchmarking, and the interested reader is referred to the websites of the CPA (http://www.cpahq.org) and the World Bank (http://www.worldbank.org), in particular, for recent developments.

References

CCAF-FCVI. 2004. "Parliamentary Oversight: Committees and Relationships." Background research paper for Review of Recent Developments in Legislative Oversight in Britain and Australia, with Special Reference to Public Accounts Committees. Ottawa: CCAF-FCVI.

CPA (Commonwealth Parliamentary Association). 2006. *Benchmarks for Democratic Legislatures: A Study Group Report*. London: CPA Secretary.

IFES (International Foundation for Electoral Systems). 2005. *Global Best Practices: A Model Annual State of the Parliament Report*. Washington, DC: IFES.

Stapenhurst, Rick, Riccardo Pelizzo, David Olson, and Lisa von Trapp. 2008. *Legislative Oversight and Budgeting: A World Perspective*. Washington, DC: World Bank.

Yamamoto, Hironori. 2007. "Tools for Parliamentary Oversight: A Comparative Study of 88 National Parliaments." Inter-Parliamentary Union, Geneva.

CHAPTER 1

Benchmarks and Self-Assessment Frameworks for Parliaments

Lisa von Trapp

"The strength of the national legislature may be a—or even the—institutional key to democratization."

—M. Steven Fish (2006)

Introduction

Elections provide a basis for rule by the people, but they do not guarantee that citizens are effectively represented. True democracy requires that those who are freely elected have the power—and the political will—to fulfill their constitutionally mandated responsibilities. Faced with challenges such as declining public confidence and executive dominance, parliaments worldwide must ensure that they function in an internally democratic manner and have the necessary authority and resources to carry out their representative, legislative, and oversight functions.[1]

Many parliaments today are seeking to improve their performance—among other things, to become more open, independent, accountable, and responsive. Every parliament is a product of its own country's history and culture, and there is no magic formula or checklist for developing an effective parliament. However, an emerging international consensus finds that certain norms and standards regarding democratic parliaments transcend the particularity of political and legislative systems. Context matters enormously, but a benchmarking or self-assessment exercise, if done well, should allow context to be fully explored.

International consensus has emerged over time on a standards-based approach in the areas of human rights and elections,[2] but until recently, a standards-based approach around what constitutes a democratic parliament had fallen behind. Today a range of international parliamentary organizations, such as the Inter-Parliamentary Union (IPU); the Commonwealth Parliamentary Association (CPA); l'Assemblée Parlementaire de la Francophonie (the Parliamentary Assembly of La Francophonie, or APF); and the Southern African Development

Community Parliamentary Forum (SADC PF), together with their partners, such as the World Bank[3] and the United Nations Development Programme (UNDP),[4] recognizes that the development of standards and assessment frameworks can (a) contribute to parliament's self-evaluative and reform efforts and (b) guide parliamentary development practitioners and donors in designing more appropriate support programs. More generally, the act of building consensus around standards is useful in further internationalizing the debate on what constitutes a democratic parliament and democracy in general.

This type of consensus building is expected to be a long-term process, and as with elections, a universal set of standards may never be agreed on. Moreover, just as there is a wide variety of organizations contributing to this work, a wide range of terminology is being used, including *standards, benchmarks, norms, criteria, indicators, principles*, and *good practices*. Nevertheless, given their shared goals and increasingly coordinated approach, the work of these different organizations has been mutually reinforcing, and a significant level of commonality can be found in the different assessment frameworks in terms of content. The reasons for this commonality are threefold:

- The frameworks have all drawn on or been influenced by one another. For example, the National Democratic Institute for International Affairs (NDI) provided technical support to the IPU as it developed its good practice guide, and the IPU guide was one of the resource documents, together with an NDI discussion paper, used by the CPA parliamentary study group to create the CPA's benchmarks for democratic legislatures.
- All of the frameworks cover the core functions of parliament—namely, approving legislation, representing citizens, overseeing the executive, and approving the budget.
- There is a common understanding of what does *not* constitute a democratic parliament, such as executive dominance and corruption of members of parliament and parliamentary officials.

Therefore, variations in vocabulary aside, it is not unthinkable that a set of overarching principles or standards for democratic parliaments could eventually be adopted.

This chapter is based on a background publication (von Trapp 2010) prepared for the International Conference on Benchmarking and Self-Assessment for Democratic Parliaments, which took place in Paris on March 2–4, 2010. The aim of this chapter is to provide a comprehensive review of the work on developing assessment frameworks for democratic parliaments that took place in the lead up to the conference. The chapter is organized as follows: First, an overview is given of the key actors and assessment frameworks. Then, a discussion of parliamentary benchmarks and self-assessment frameworks as part of a larger trend follows. Commonalities and key differences across the frameworks are reviewed, and initial guidelines are suggested for using the frameworks and experiences at the national and state level. The chapter closes with some concluding remarks.

Key Actors and Assessment Frameworks

The main actors in the process of developing standards for democratic legislatures are organizations or associations of parliaments or parliamentarians. These organizations represent a broad spectrum of parliaments and parliamentarians from across the globe and are able to bring their members' views to bear in the discussion. They are well placed to understand both the shared traits and the diversity represented in parliamentary democracies today. Other actors, such as the World Bank and UNDP, play a supportive role by mobilizing resources and providing expertise as appropriate. UNDP, the World Bank, NDI, and others bring valuable experience from their own parliamentary strengthening work with a range of legislatures.

The frameworks described in this section are living documents or works in progress that are open to eventual adaptation and elaboration. As they are discussed internationally and regionally and as they are applied at the national level, they will change. Indeed, specific regional considerations have already been identified, and new benchmarks have been suggested during the SADC PF workshops and during the CPA regional workshops for the Pacific and Asia. Moreover, just as legislatures are continuously evolving, standards will likely evolve and presumably rise. In the future, some organizations may even choose to develop more aspirational benchmarks.[5]

This section outlines four of the most commonly cited frameworks: (a) the NDI's International Standards for Democratic Legislatures, (b) the CPA's Recommended Benchmarks for Democratic Legislatures, (c) the APF's *critères d'évaluation*, and (d) the IPU's Self-Assessment Toolkit for Parliaments. For further information on the historical debate on what constitutes a democratic parliament, refer to annex 1A.

The NDI's International Standards for Democratic Legislatures

Founded in 1983, NDI describes itself as "a nonprofit, nonpartisan organization working to support and strengthen democratic institutions worldwide through citizen participation, openness, and accountability in government."[6] NDI has worked with individual members, parliamentary leaderships, committees, and political party caucuses in national and regional legislatures in more than 60 countries.

In January 2007, NDI published a discussion document titled "Toward the Development of International Standards for Democratic Legislatures" (NDI 2007). The process leading up to this publication involved a broad survey of existing documents from a range of organizations, such as the IPU, CPA, Organization for Security and Co-operation in Europe, Organisation for Economic Co-operation and Development, SADC, International Conference of New or Restored Democracies, Community of Democracies, and United Nations. Thus, the 88 standards identified in the NDI publication represent an attempt to codify already widely agreed principles. The standards are grouped into four main categories: (a) election and status of legislators,

(b) organization of the legislature, (c) functions of the legislature, and (d) values of the legislature. Box 1B.1 in annex 1B provides an overview of the standards within these categories.

In 2008–09, NDI also designed a survey tool, the Standards-Based Questionnaire, which attempts to determine perceptions of the legislature's (formal) authority and of its performance (that is, its behavior in practice). The survey covers 25 issues that are often included in benchmarks for democratic parliaments or in parliamentary self-assessment tools.

The CPA's Recommended Benchmarks for Democratic Legislatures
Established in 1911, the CPA is an "association of Commonwealth parliamentarians who, irrespective of gender, race, religion, or culture, are united by community of interest, respect for the rule of law and individual rights and freedoms, and pursuit of the positive ideals of parliamentary democracy."[7] Through a variety of activities, the CPA seeks to "promote the advancement of parliamentary democracy, to build an informed parliamentary community able to defend the Commonwealth democratic commitment, and to further cooperation among its member Parliaments and legislatures." The CPA's membership comprises around 17,000 parliamentarians from around 175 national, state, provincial, and territorial parliaments in Commonwealth countries.[8]

In November 2006, the CPA convened the Parliamentary Study Group, with CPA members representing different Commonwealth regions.[9] Building on the Commonwealth (Latimer House) Principles on the Accountability of and Relationship between the Three Branches of Government (CPA and others 2004), the NDI discussion paper, and the recommendations of 26 previous CPA workshops and study groups,[10] the group worked to synthesize and codify a set of benchmarks to reflect the current state of good Commonwealth parliamentary practice. The group considered the following themes and recommended a set of benchmarks related to each:

- Representative aspects of parliament
- Assurance of the independence, effectiveness, and accountability of parliament
- Parliamentary procedures
- Public accountability
- Parliamentary service
- Parliament and the media

The end product is a set of 87 benchmarks that attempt to cover the features of a "fully functioning and empowered democratic parliament" (von Trapp 2007). As explained in chapter 3, these benchmarks are divided into four main topical headings: general, organization of the legislature, functions of the legislature, and values of the legislature. The CPA benchmarks are phrased as statements rather than questions, and no specific system or methodology to code responses to these benchmarks is provided. A CPA guidance note explains that the benchmarks are a useful tool to launch a debate, to provide a basis for measuring parliamentary effectiveness, or to help leverage reforms (CPA 2009).

According to CPA practice, the benchmarks are intended to be the beginning of a larger discussion rather than an end in themselves. The CPA expects that the development of regional versions of the benchmarks that reflect the diverse practices and priorities within the Commonwealth will also contribute to the evolution of the benchmarks. Some CPA regions have developed their own versions of the benchmarks, which affirm the majority of the original benchmarks while adding several new benchmarks. At the same time, the CPA is encouraging individual parliaments to undertake benchmark self-assessments and to share their experiences with their peers in other Commonwealth parliaments. Benchmarking assessments have been conducted by the Australian Capital Territory (see chapter 12), Bermuda (see chapter 14), Canada (see chapter 11), Kiribati (see chapter 13), Nauru, Niue, and Tuvalu, among others.

The APF's Critères d'Évaluation

Established in Luxembourg in 1967, the APF is the consultative assembly of the Organisation Internationale de la Francophonie (International Organization of La Francophonie, or OIF). The APF brings together 77 parliaments from four geographic regions: Africa, the Americas, Asia-Pacific, and Europe. It works through four standing committees and a network of women parliamentarians. The APF seeks to promote democracy, peace, and human rights; to enhance the influence of parliamentarians; and to promote the French language.[11] It provides technical assistance to member parliaments and is currently collecting a compendium of parliamentary procedures and practice from its members. The development of standards, in partnership with UNDP, is therefore seen as a natural extension of the APF's core mission.

The APF took the CPA benchmarks as a starting point in developing a set of parliamentary standards. It also conducted a comparative study of the rules of procedure of several parliaments within the francophone countries, and drew on the work of the OIF.[12] Although many of the APF criteria match (or are similar to) the CPA benchmarks, the APF has gone further in some areas. For example, the APF has developed additional criteria around elections or measures to be included in parliaments' rules of procedure and has significantly expanded the number of benchmarks on participation in international affairs. The final result of the APF's exercise is 117 criteria, which were formally and unanimously adopted as "La réalité démocratique des Parlements: Quels critères d'évaluation?" ("The Reality of Democratic Parliaments: What Criteria of Evaluation?") during the 35th session of the APF in Paris on July 6, 2009. Chapter 5 considers this document (APF 2009) in more detail.

The IPU's Self-Assessment Toolkit for Parliaments

Established in 1889, the IPU is the world's oldest interparliamentary organization. The IPU has more than 160 national parliaments as members and 10 regional parliaments as associate members. Members are divided into six geopolitical groups: Africa, the Arab Group, Asia-Pacific, Eurasia, Latin America, and Twelve Plus. However, some IPU members are not affiliated with any of these geopolitical groups. The IPU's main areas of activity are representative democracy; human rights and humanitarian law; international peace and security; women in politics; sustainable development; and education, science, and culture.[13]

Published in 2008, the IPU toolkit developed out of a "major programme of work undertaken by the IPU, to examine what makes a parliament *democratic*, both in the way it functions and interacts with its electorate, and in its effectiveness in performing its roles within a democratic system of government" (IPU 2009b, 1, emphasis in original). The toolkit builds on a collection of best practices from many of the organization's member parliaments, as well as on consultations with an expert working group. The toolkit's self-assessment methodology draws extensively from the State of Democracy Assessment Methodology developed by the International Institute for Democracy and Electoral Assistance (International IDEA).

The main objective of the IPU self-assessment toolkit is to assist parliaments in a systematic analysis of their performance, to identify their strengths and weaknesses, and to formulate recommendations for reform and development. The IPU self-assessment toolkit comprises 54 questions organized around six categories (IPU 2008, 5):

• The representativeness of parliament
• Parliamentary oversight over the executive
• The parliament's legislative capacity
• The transparency and accessibility of parliament
• The accountability of parliament
• The parliament's involvement in international policy

The toolkit is considered in more detail in chapter 2.

The IPU toolkit questions are framed in the comparative mode and ask how effective, adequate, systematic, and so forth the parliament is. A five-point scale is used to measure responses. Further questions then ask for the biggest recent improvement in each respective section, the most serious ongoing deficiency, and potential measures to remedy this problem.

The toolkit suggests a number of possible scenarios for its use but notes that the precise format for using the toolkit will depend on its purpose (IPU 2008, 12). The IPU believes that parliamentarians themselves are best placed to identify the challenges they face in practice and to suggest ways in which those problems may be overcome. Thus, the initiative for self-assessment should come from parliaments themselves. The toolkit suggests that key parliamentarians should be involved and that the assessment group should reflect the broadest possible range of perspectives from within the parliament. Some parliaments may choose to work in partnership with an outside organization or with outside experts or facilitators. In that case, participants should agree on the precise role and scope of the exercise in advance, as well as on the expected timescale and outcomes of the process (IPU 2009b, 2).

The IPU has trained facilitators to assist in the assessment process as requested. As a result of lessons learned during a first round of self-assessments in Rwanda, Sierra Leone, and several other countries, the IPU drafted a preparation note for parliaments on carrying out a self-assessment to provide further guidance (IPU 2009a). Promoting the toolkit remains a high priority, and the IPU has initiated discussions with regional parliamentary organizations with a view to increasing awareness and use of the toolkit.

Parliamentary Benchmarks and Self-Assessment Frameworks as Part of a Larger Trend

Several other organizations, such as the Canadian Parliamentary Centre, have also developed parliamentary assessment frameworks. Others, such as the Parliamentary Assembly of the Council of Europe (PACE) have expressed an interest in developing their own evaluative frameworks. Still others are including components on parliament within broader assessment frameworks such as the International IDEA State of Democracy Assessment Methodology or the Transparency International (TI) National Integrity System Assessment. Certain benchmarks or standards around democratic parliaments are reflected in sets of governance indicators such as the World Bank's actionable governance indicators (AGIs). This section illustrates a few of these examples.[14]

The Parliamentary Centre's Parliamentary Report Card Methodology

The Parliamentary Centre has developed the Parliamentary Report Card methodology (see figures 1.1 and 1.2) and a related set of 37 indicators on the budget process. The Parliamentary Report Card tests performance in four areas that are almost universally regarded as the core functions of parliament: legislation, representation, oversight, and the budget. It then evaluates these four lines of service against five performance tests on the level and range of activity, openness and transparency, participation, accountability, and policy and program impact.

Figure 1.1 Parliamentary Report Card

		Legislation	Oversight	Representation	Budget
Performance tests	Level and range of activity				
	Openness and transparency				
	Participation				
	Accountability				
	Policy and program impact				

Source: Parliamentary Centre, Canada.

Figure 1.2 Sample of Report Card Performance Area and Related Indicators

Accountability	• Does the parliament have a public accounts committee or an equivalent entity that examines past expenditures?
	• Are measures taken to ensure its independence such as the appointment of an opposition member of parliament as the chair?
	• Does the public accounts committee work with independent audit authorities to uncover financial irregularities and promote program efficiency?
	• Does the parliament question government leaders, ministers, and officials fully during the budget process?
	• Does the parliament effectively scrutinize departmental work plans and monitor their implementation?
	• Does the parliament undertake program and policy evaluations?
	• Does the parliament review commitments entered into by senior public servants?

Source: Parliamentary Centre, Canada.

The indicators are phrased as questions, and respondents use a scale of zero to five. The Parliamentary Centre has begun limited field-testing of the Parliamentary Report Card using this first set of indicators in Cambodia and in several African countries (for the latter, see chapters 9 and 15). To date, the methodology has not been widely used, and work is in progress to refine the methodology and to develop new indicators to better inform their assistance programs. Figures 1.1 and 1.2 demonstrate the format of the report card itself and sample indicators.

The Parliamentary Assembly of the Council of Europe and Self-Evaluation

PACE was the first European regional parliamentary group to explore assessment frameworks. In January 2009, PACE's Bureau of the Assembly referred a motion to consider elaborating procedural guidelines for self-evaluation by national parliaments in Europe. The Committee on Rules of Procedure, Immunities, and Institutional Affairs has since produced and debated a draft paper titled "Self-Evaluation by Europe's National Parliaments: Procedural Guidelines," which takes into account work by the IPU and other organizations (PACE 2009). The committee then began work on a questionnaire for PACE's members and considered several follow-up steps in this workstream, including (a) analyzing the relevance of existing self-assessment standards in the parliaments of Council of Europe (CoE) member states, (b) providing information on the strengths and weaknesses of CoE parliaments and identifying a model for an exemplary parliament, and (c) discussing the appropriateness of procedural guidelines for performance assessment by international parliamentary institutions in Europe.

International IDEA's State of Democracy Assessment Methodology

As noted earlier, the IPU's (2008) Self-Assessment Toolkit for Parliaments draws extensively from International IDEA's State of Democracy Assessment Methodology. The IDEA methodology is a reform-oriented assessment with several aims: (a) to generate debate among stakeholders on various issues identified by the assessment, (b) to feed into evidence-based advocacy, (c) to contribute to policy reform, and (d) to raise awareness about the quality of democracy in the country assessed.[15] International IDEA's assessment framework has 4 pillars and 15 subpillars, each of which is assessed by answering a series of questions that examine whether certain democratic institutions and processes are in place and how they perform in practice. One of the subpillars focuses on the democratic effectiveness of parliament.

Transparency International's National Integrity System Assessment

In 2009, Transparency International devised new indicators for the pillar "legislature," one of 12 institutions assessed by TI's National Integrity System.[16] The purpose of this pillar was to examine the different parliamentary benchmarks and self-assessment frameworks. Similar to NDI's Standards-Based Questionnaire, the TI tool indicators attempt to measure both formal powers (law) and practice.

World Bank's Actionable Governance Indicators

The World Bank's AGIs—described as "narrowly defined and disaggregated indicators that focus on relatively specific aspects of governance and could provide guidance on the design of reforms and monitoring of impacts"—reflect several of the standards identified by parliamentary organizations (Reid 2008).[17] For example, Public Expenditure and Financial Accountability indicators PI-27 ("Legislative Scrutiny of Annual Budget Law") and PI-28 ("Legislative Scrutiny of External Audit Reports") are directly related to the standards addressing parliament's role in the budget process. Human Resource Management (HRM) Performance Indicators and the HRM Diagnostic Instrument also contain indicators that can be linked to standards dealing with recruitment, retention, and codes of conduct for parliamentary staff members. Furthermore, other multilateral initiatives, such as the Global Initiative on Fiscal Transparency and the Open Government Partnership, all have components that relate to the role of parliament in the budget process or transparency.

The Frameworks: Commonalities and Differences

With this background, it is useful to take a closer look at the commonalities and differences across the various parliamentary assessment frameworks. To facilitate this examination, annex 1D provides a comparison table using the CPA benchmarks as a starting point and comparing them to the NDI standards and the APF criteria. Although the CPA benchmarks are presented in order, the NDI standards and APF criteria are not; instead they are presented in relation to the CPA benchmarks. The table uses a color-coded system. Benchmarks, standards, or criteria that match are coded as light gray, those that are very similar are coded as medium gray, and those that are new or that only appear in one set of standards are coded as dark gray. For reference, the Participants' Statement from the March 2010 International Conference on Benchmarking and Self-Assessment for

Democratic Legislatures also provides a brief summary of areas of consensus under five themes: institutional independence, procedural fairness, democratic legitimacy and representation, parliamentary organization, and core legislative and oversight functions.[18]

The comparative table in annex 1D allows readers to see the high level of consensus that exists between the main frameworks. Although there are differences between the frameworks, it quickly becomes apparent that those differences do not stem from conflicting principles but from different areas of focus or parliamentary traditions. For example, greater emphasis is given to ex post financial oversight and the specific role of public accounts committees in the CPA and SADC PF benchmarks,[19] no doubt because of their importance in most Westminster-based systems.

The IPU's toolkit does not lend itself to this type of comparison table, but certain questions can be matched to the different standards, and many of the possible "procedural and institutional means" identified in the IPU Framework on the Parliamentary Contribution to Democracy mirror the recommended benchmarks. For example, the framework identifies parliament's control of its own budget, a nonpartisan professional staff separate from the main civil service, and adequate unbiased research and information facilities for members as procedural and institutional means for ensuring parliamentary effectiveness. All three of these issues are covered by the CPA benchmarks, NDI standards, and APF criteria.

The table in annex 1D reveals that more than 80 percent of the CPA benchmarks and the NDI standards are the same or similar. Differences include NDI standards (some of which are arguably influenced by the U.S. experience) that legislators have the right in nonparty-list electoral systems to leave their party group (4.2.2); that no partisan or nonpartisan staff member shall have any legislative or procedural authority, including voting (5.3.2); that the legislature shall have the power to amend the budget (6.3.1); that in the absence of a public referendum, constitutional amendments require the legislature's approval (6.5.1); that the legislature have a nonpartisan ombudsman or similar body that investigates complaints of executive branch malfeasance and makes recommendations and reports directly to the legislature (7.3.1); that the legislature's consent be required in the confirmation of senior judges and the legislature shall have mechanisms to impeach judges for serious crimes (7.6.1); and that the legislature be accessible to persons with disabilities (9.2.3).

The vast majority of both the CPA benchmarks and NDI standards also match or are similar to the APF criteria. However, the APF has developed around 30 additional criteria. In some cases, the APF simply uses two criteria to address what the CPA combines in one benchmark, or vice versa. Often new APF criteria specify that certain aspects be defined in the constitution, by law, or in the rules of procedure. This development is interesting, as other groups have shied away from benchmarks that force specific changes to the constitution or rules of procedure.

The APF has also developed additional criteria around topics such as elections and has significantly expanded the number of benchmarks on parliamentarians' and parliaments' participation in international affairs. For example, criterion 2.5.2.3 calls for parliamentarians to be included in government delegations participating in international negotiations. Another significant difference from the NDI and CPA frameworks is that the APF has not adopted criteria on no-confidence and impeachment measures. Finally, the APF criteria contain specific benchmarks related to gender equality that are not found in the CPA or NDI frameworks (such as 3.2.1.5, which calls for representation of women at all levels of the parliamentary administration, and 2.1.1.3, which calls on parliaments to preserve a balanced representation of women and men at all levels of responsibility within parliament).

Despite these variations, many areas of consensus across the different standards remain. For example, all frameworks recognize the right of legislators to immunity for speech during the exercise of their duties. Moreover, to reinforce the autonomy of the legislature, all recognize that the executive branch shall have no right or power to lift the immunity of a legislator.[20]

Other measures to increase parliament's autonomy that are addressed by the different standards include providing proper remuneration and reimbursement of expenses to legislators, as well as adequate physical infrastructure, information and communication technology facilities, and nonpartisan professional staff support. In addition, there are standards on the legislature's control of the parliamentary service and terms of employment, including that the legislature have adequate resources to recruit a staff sufficient to fulfill its responsibilities, that the rates of pay for the parliamentary staff be broadly comparable to those of the civil service, and that recruitment be based on merit.[21] These standards in part seek to address concerns that qualified staff members may be deterred from staying in the parliamentary service because of lower pay and benefits. Moreover, as the parliamentary service is part of the civil service more generally and is controlled by the executive, there is a risk that staff could be moved to other areas of the civil service, potentially disrupting the work of parliament.[22] A final concern is that staff members who assist committees in conducting inquiries may feel pressured to tone down resulting reports if they reflect negatively on the executive.

Although the funding models differ, all frameworks recognize that a legislature's ability to determine and approve its own budget is essential to ensuring its independence. For the legislature to exercise oversight of the executive branch, the legislature's budget must not depend on the executive branch. This idea is consistent with additional standards recommending that the approval of the legislature be required for the passage of all legislation, including budgets.

There is also broad consensus that only parliament may adopt and amend its rules of procedure. Similarly, there is agreement that the legislature should meet regularly at intervals sufficient to fulfill its responsibilities and that the legislature should have procedures for calling itself into extraordinary or special session. In addition, the different standards cover legislatures' right to amend proposed agendas for debate.

Another area of broad consensus is the legislature's right to form permanent and temporary committees, the presumption that the legislature will refer legislation to a committee, and the right of the committee to amend legislation referred to it.[23] Committees—often described as the "engine room" of the legislature—have emerged as among the most critical tools at legislatures' disposal today. Committees allow for more in-depth scrutiny and, particularly through holding hearings, provide an important avenue for public input.[24] Working in a committee allows legislators to develop specialized knowledge on matters within the jurisdiction of their committee. Lastly, work in a committee is often viewed as less partisan.[25]

In terms of powers, the standards also address committees' right to summon persons, papers, and records and right to consult or employ experts. In terms of organization, the standards call for committee membership to reflect the political composition of the legislature.[26]

Broadly agreed standards related to transparency include standards that votes be public, that the legislature publish records of its proceedings, and that the legislature be accessible to citizens and to the media.[27] Matters of transparency and integrity are also addressed through, for example, standards on public disclosure of financial assets and business interests; requirements that there be mechanisms to prevent, detect, and bring to justice legislators and staff members who are engaged in corrupt practices; and, in the case of the CPA and NDI, standards on codes of conduct for legislators and parliamentary staff members.

Differences across frameworks can also be found at the regional level. Regional benchmark discussions have affirmed existing benchmarks and standards, while developing new—and sometimes regionally specific—benchmarks. Annex 1E summarizes regional trends in parliamentary assessment frameworks.

The Frameworks and Gender

Gender-related concerns have been incorporated into the discussion and development of the different assessment frameworks from the outset. Women parliamentarians composed almost half of the original study group for the CPA Recommended Benchmarks for Democratic Legislatures, and feedback on the benchmarks was sought from the Commonwealth Women Parliamentarians Steering Committee. The APF asked its Women Parliamentarians Network to review its draft benchmarks, and SADC PF has undertaken a similar process through its Standing Committee on Democratization, Governance, and Gender Equality and Regional Women's Parliamentary Caucus. The IPU toolkit includes promoting gender sensitivity in parliament as one of the scenarios for use and provides questions to stir dialogue and debate. For example, question

3.7 asks, "How careful is parliament in ensuring a gender-equality perspective in its work?".

Specific benchmarks prohibit restrictions on candidate eligibility that are based on gender, and SADC PF addresses the question of representation through several additional benchmarks (SADC PF 2010). For example, it includes these benchmarks:

- Parliamentary membership shall reflect the social diversity of the population with respect to gender, language, religion, and ethnicity among other considerations.
- Parliaments shall enact laws that require political parties to take measures of affirmative action for gender to meet the provisions of the Southern African Development Community Protocol on Gender and Development.
- Nomination fees must be reasonable and affordable so as not to unduly exclude potential candidates.

The APF criteria and SADC PF benchmarks address the need for gender balance in the parliamentary leadership. APF criterion 2.1.1.3 states, "Le Parlement doit prendre des mesures significatives visant à établir et préserver une proportion équilibrée de femmes et d'hommes dans ses différentes instances à tous les niveaux de responsabilité" ("The legislature shall take significant steps to establish and preserve gender balance in its various bodies at all level of responsibility"). According to benchmark 5.7.3, "There shall be equitable gender representation in the election of presiding officers." Similarly, both call for gender to be taken into consideration in the composition of parliamentary committees.

Bearing in mind the importance of women's caucuses in many parliaments around the world, the CPA, NDI, and SADC PF include benchmarks regarding parliamentarians' right to form cross-party caucuses (although the CPA's Asia, India, and Southeast Asia regions removed this benchmark in the regional version). The APF also calls for gender to be taken into account in the composition of any official parliamentary delegations. Finally, all three standards contain benchmarks that prohibit discrimination based on gender in the recruitment and promotion of parliamentary staff members, and the APF calls explicitly for the representation of women at all levels in the parliamentary staff hierarchy.

Initial Guidelines for Using the Frameworks and Experiences at the National Level

Entry Points for Benchmarking and Self-Assessment Exercises

There are several entry points for use of the different assessment frameworks. As noted in the introduction to the IPU toolkit, all of the scenarios for self-assessment share two objectives: (a) "to evaluate Parliament against international criteria for democratic Parliaments" and (b) "to identify priorities and means for strengthening Parliament" (IPU 2008). As such, the frameworks can have the following purposes:

- They can help prepare the parliamentary budget or strategic plan.
- They can stimulate a parliamentary reform process.
- They can promote debate.
- They can enable new members of parliament to discuss key issues.
- They can conduct a review or create a baseline for measuring progress.
- They can validate the findings of a needs-assessment mission.
- They can support a request for external assistance.
- They can make a civil society organization assessment of parliament.
- They can promote gender sensitivity in parliament.

NDI also identifies several similar uses for its questionnaire: as a diagnostic tool to help identify priorities for legislative strengthening work (a point worth noting for donors), as an advocacy tool to support parliamentary reform, and as the basis for dialogue between parliamentarians and civil society representatives (NDI 2009, 2).

Peer review mechanisms may also find these types of standards or benchmarks helpful, as evidenced by the South Africa case study in which the assessment of the South African parliament, which drew on the work of the IPU and the World Bank among others, was originally conceived as part of the parliament's engagement with South Africa's African Peer Review Mechanism process.

Emerging Methodologies for Use

The range of parliamentary frameworks allows for a great deal of flexibility in how they are used. This flexibility is an advantage in that it allows parliaments to adapt the frameworks to their specific needs. However, as the frameworks have begun to be applied at the country level, a need for additional methodological guidance has become apparent. Given the recent nature of this work, there are few national case studies to date, and because the frameworks have been applied differently in different countries, drawing strong comparisons or identifying lessons learned is difficult at this stage. Moreover, even in cases where parliaments have used one of the assessment frameworks and made recommendations for reform, we have yet to see how successful they will be in following up on and implementing their recommendations over time.

Although the frameworks have been developed to be as universally relevant as possible, the different benchmarks and standards are seen as a starting point that may need refinement and adaptation at the regional level. Depending on the country and that country's particular practices, some standards may be more relevant than others. Similarly, the IPU's toolkit has been designed as a generic document that can be used in many circumstances; however, depending on the country using it, some questions may be more or less relevant and require amendments.

The IPU has arguably gone furthest in developing a methodology, which, as noted earlier, is based on International IDEA's State of Democracy methodology. Section II of the IPU toolkit contains suggestions on how to use the toolkit—namely, ways to initiate a self-assessment, potential participants in the process, questions to include, the role of facilitators, ways to set a timeframe for the self-assessment, additional sources of data, ways to document the process, expected outcomes, and issues surrounding publicity. On the basis of its initial experiences with countries using the toolkit, the IPU also developed "Carrying Out a Self-Assessment: Preparation Note for Parliaments" (IPU 2009a). This document identifies nine steps for carrying out a self-assessment that could easily be applied to a benchmarking exercise.

The CPA also recently prepared a guidance note (CPA 2009). Although the CPA benchmarks are phrased as standards rather than as questions and no system to code or categorize responses to these benchmarks is provided, the note explains that a "benchmarks" self-assessment can be conducted at several levels (CPA 2009, 3):

- At the most basic level, the self-assessment can determine whether the parliament or legislature is able to "tick the box" on each of the 87 standards.
- It can also assess how well the parliament meets each standard—for example, by rating the standard on a scale of one to five.
- The self-assessment can devise another method of scoring, such as by setting top scores higher in the areas considered most important.

The CPA note also provides suggestions for the composition of a benchmarks self-assessment panel (presiding officers, government and opposition members, clerks or secretaries, and other officials) and notes that the panel may have added credibility if it includes some respected and knowledgeable external assessors (such as judges, senior civil servants, lawyers, academics, or former members or officials).

As noted earlier, NDI has developed a separate questionnaire that is based on 25 standards that it has identified. For each of these 25 issues, the questionnaire contains two related statements. The first focuses on the formal powers of the legislature, and the second relates to whether the power is used in practice. Survey participants are asked to indicate whether they strongly agree, agree, disagree, or strongly disagree with each of these 25 two-part statements. Participants may indicate that they are unaware of the issue or that the issue does not apply to their parliament. The survey should be administered to parliamentarians, parliamentary staff members, and representatives of civil society to allow their perceptions to be compared. NDI may also disaggregate survey responses by gender or by other factors, such as length of service in parliament.

In practice, NDI has mainly administered the survey as part of multiday training sessions, with surveys being collected on the first day of the workshop and preliminary analysis of the results being presented on the second and third days. Although this approach may limit the sample size, NDI notes that participants have generally found the initial anecdotal data from the surveys useful for launching discussions around potential explanations for the survey results.[28] NDI continues to look at refining the survey tool and its methodology (NDI 2009, 3).

Conclusions

It is critical that parliaments and parliamentarians be engaged in identifying the criteria that they feel are important to a democratic parliament. Several interparliamentary organizations (representing tens of thousands of parliamentarians worldwide) and their partners have undertaken a significant program of work in developing assessment frameworks for democratic parliaments. All of the tools look to some degree at legislatures' core functions (that is, their representative, legislative, and oversight functions). All frameworks also place great emphasis on accountability, participation, openness, and transparency. As such, there is significant overlap among the different frameworks in terms of content, particularly between the different versions of the standards, benchmarks, and criteria. Annex 1D allows readers for the first time to see, in a simple and practical way, the broad areas of consensus among the major frameworks in existence today.

Consensus building is expected to be a long-term process, and as with elections, there may never be a universally agreed-on set of standards. While supporting a pluralistic approach, the different organizations involved are nevertheless already working in a coordinated, mutually reinforcing manner, and initial feedback at the regional level affirms the frameworks' relevance. Ideally, further regional interparliamentary organizations and associations will join the debate. Just as legislatures are continuously evolving, standards will likely evolve and presumably rise. Indeed, while some parliaments may find the current standards difficult to meet, others see them as not going far enough. In the future, some interparliamentary organizations may even choose to develop more aspirational benchmarks.

Parliaments should not be discouraged if they do not meet a specific benchmark; rather they should take the opportunity to debate the principle behind the benchmark, to discuss the relevance of the benchmark to their specific context, and to determine whether they would benefit from changes that would allow them to meet the benchmark.

Parliaments are just beginning to test or apply the different frameworks and to provide feedback on their experiences. Many will need assistance from partners in the parliamentary strengthening and donor community to take such an exercise forward. Only as (or if) the frameworks become more widely used will they truly be owned by parliaments themselves. Moreover, only parliaments' use of these frameworks will validate the frameworks' relevance and applicability and will reveal lessons for modification or adaptation. This point is extremely important given the competing demands on parliaments. In addition, the strength of individual benchmarking or self-assessment exercises lies in the willingness of a given parliament and its members to engage with the issues. Assessments that are part of a larger process of relevance to the parliament, in which the results are followed up, are preferable to one-off exercises.

Annex 1A: Overview of the Process, 2004–10

Although elements of what constitutes a democratic parliament have been discussed for some time, 2004 marked the beginning of a more structured, multiactor process to address this issue with zest. In December 2004, representatives from 15 organizations came together to enhance collaboration on this topic at a CPA–World Bank organized workshop, Parliamentary Standards for Democratic Legislatures, in Washington, DC.[29] Subsequently, a series of study groups, workshops, and other forums have allowed various organizations to make significant advances in this domain.

From 2006 to 2009, NDI developed a suggested set of minimum standards for democratic legislatures, as well as a standards-based questionnaire, which attempts to determine perceptions of both a legislature's formal powers and actual practice. The questionnaire has been tested in several countries in Latin America and the Balkan states.

A 2006 CPA parliamentary study group produced a set of Recommended Benchmarks for Democratic Legislatures (CPA 2006). The benchmarks are currently being examined and adapted at the regional level in the CPA's Pacific and Asian regions (Asia, India, and Southeast Asia) and several countries, including Canada and some Pacific countries, have tested the benchmarks at the national level. Using the CPA benchmarks as a starting point, given that their membership overlaps broadly, SADC PF drafted a first set of regional benchmarks during two workshops in 2007 and 2009, respectively. Prior to the workshops, SADC PF also hired two consultants to research best practices in the region and relevant constitutional and legal frameworks.

The APF also used the CPA benchmarks as a foundation for the development of criteria to suit the traditions and practices of francophone parliaments. Working through its Political Affairs Commission, Parliamentary Affairs Commission, and network of women parliamentarians, the APF developed and then adopted "La réalité démocratique des Parlements: Quels critères d'évaluation?" (APF 2009) during its annual assembly in Paris in July 2009. The APF criteria were disseminated to APF member parliaments and the Association des Secrétaires Généraux des Parlements Francophones (Association of Secretaries General of Francophone Parliaments), among others.

The IPU has collected examples of good practice from 75, or around half, of its member parliaments. These examples formed the basis for the 2006 IPU publication *Parliament and Democracy in the Twenty-First Century: A Guide to Good Practice* (Beetham 2006) and the 2008 Self-Assessment Toolkit for Parliaments (IPU 2008). Although its approach may appear somewhat different from that of its partners, the IPU too sought to base its toolkit on "universal democratic values and principles … relevant to all parliaments, whatever political system they adhere to, whatever their stage of development" (IPU 2008, 5). The toolkit was presented to IPU members during a special workshop at the IPU's 2008 Annual Assembly and has been used by the Cambodian Senate, a Pakistani think tank, the parliaments of Rwanda and Sierra Leone, and the institutional performance task team in the South African parliament. Assessments are also under way for the parliaments of Andorra and Ireland. In October 2009, the IPU and the Association of Secretaries General of Parliaments (ASGP) convened a one-day conference with partners, including the CPA, NDI, and the APF. The conference was titled Evaluating Parliament: Objectives, Methods, Results, and Impact. On the basis of trials conducted by ASGP staff members in Algeria and Sri Lanka, the ASGP began work on a similar toolkit targeted at the parliamentary administration.

In addition to these efforts, the World Bank and Griffith University convened a smaller international workshop on legislative benchmarks and indicators in Brisbane, Australia, in 2008. The workshop brought together several representatives of the same organizations with other legislative development practitioners, academics, and civil society organization representatives to discuss ways to assess legislative performance and the effectiveness of legislative strengthening programs. At the close of the workshop, participants identified a series of steps to take this work forward:

- Create a small steering committee to coordinate efforts.[30]
- Promote a research agenda to apply the different frameworks at the country level.
- Hold a larger international conference in early 2010 to take stock of developments, identify areas of broad consensus, and begin to draw lessons of experience from applications at the national level.

After the Brisbane workshop, this work was presented in several larger forums, including a Wilton Park conference titled Enhancing the Effectiveness of Parliaments and the Second Donor Coordination Meeting on Parliamentary Development, both held in October 2008.

The March 2010 International Conference on Benchmarking and Self-Assessment for Democratic Legislatures in Paris was a result of the Brisbane meeting and the work of the steering committee.[31] The conference objectives included

- Identifying areas of internationally agreed consensus among the current sets of standards and principles, as well as areas of potential further consensus
- Examining national case studies and drawing a first set of lessons of experience
- Broadening the research agenda and encouraging future applications of the different frameworks at the national level
- Bringing in regional perspectives to the dialogue on standards for democratic parliaments

- Inviting additional regional interparliamentary organizations to join the process to ensure broad representation and ownership.

The conference closed with participants agreeing on a statement with recommendations for parliaments, parliamentary strengthening organizations, donors, and other actors.

Box 1A.1 provides a timeline of the process events.

Box 1A.1 Process Events Timeline

2003
- The Parliamentary Centre and the World Bank develop the Parliamentary Report Card methodology and related indicators of parliamentary performance in the budget process.

2004
- September: The Commonwealth Parliamentary Association (CPA) holds a panel discussion on its Recommended Benchmarks for Democratic Legislatures during its 50th Annual Conference in Quebec and Ontario.
- December: The World Bank and the CPA host a workshop, Parliamentary Standards for Democratic Legislatures, in Washington, DC.

2006
- The Inter-Parliamentary Union (IPU) publishes *Parliament and Democracy in the Twenty-First Century: A Guide to Good Practice* (Beetham 2006).
- The National Democratic Institute of International Affairs (NDI) disseminates the first draft of a discussion document titled "Toward the Development of International Standards for Democratic Legislatures."
- October: The CPA holds a parliamentary study group in Bermuda on benchmarks for democratic legislatures.
- December: The CPA publishes "Recommended Benchmarks for Democratic Legislatures" (CPA 2006).

2007
- January: NDI publishes "Toward the Development of International Standards for Democratic Legislatures" (NDI 2007).
- May: The U.K. Department for International Development (DFID), United Nations Development Programme (UNDP), and World Bank hold the First Donor Consultation on Parliamentary Development and Financial Accountability in Brussels.
- September: The CPA holds a panel discussion on the CPA benchmarks during its 53rd Annual Conference, which convenes in New Delhi.
- November: The Southern African Development Community Parliamentary Forum holds a workshop in Pretoria titled Benchmarks for Democratic Legislatures in Southern Africa.

box continues next page

Box 1A.1 Process Events Timeline *(continued)*

2008

- July: L'Assemblée Parlementaire de la Francophonie (the Parliamentary Assembly of La Francophonie, or APF) begins its benchmarks process.
- September: The IPU publishes its Self-Assessment Toolkit for Parliaments (IPU 2008). NDI develops the first draft of its minimum standards assessment survey and tests it in the Balkans. The World Bank and Griffith University hold a workshop on legislative benchmarks and indicators in Brisbane, Australia, where an informal steering committee is formed.
- October: The IPU holds a workshop on self-assessment during its Annual Assembly in Geneva. Partners participate in a Wilton Park conference titled Enhancing the Effectiveness of Parliaments. The DFID, UNDP, and World Bank hold the Second Donor Coordination Meeting on Parliamentary Development in London.
- December: The IPU holds a training workshop for facilitators in Geneva on the use of the IPU's Self-Assessment Toolkit for Parliaments.

2009

- March: The IPU drafts "Carrying Out a Self-Assessment: Preparation Note for Parliaments" (IPU 2009a). The APF holds a seminar in Fribourg, Switzerland, on the synthesis of the APF criteria. Using the IPU toolkit, the Pakistan Institute of Legislative Development and Transparency publishes "State of Democracy in Pakistan: Evaluation of the Parliament 2008–2009" (PILDAT 2009).
- April: The APF holds a seminar in Luang Prabang, Lao People's Democratic Republic, on the synthesis of the APF criteria.
- June: The CPA, World Bank, and Centre for Democratic Institutions (CDI) hold a first workshop in Brisbane, Australia, titled Pacific Benchmarks for Democratic Legislatures, following a CDI professional development course for parliamentary speakers from Pacific island countries.
- July: The APF Annual Assembly, held in Paris, adopts "La réalité démocratique des Parlements: Quels critères d'évaluation" (APF 2009).
- September: The CPA drafts "CPA Benchmarks for Democratic Legislatures: Self-Assessment Guidance Note" (CPA 2009).
- October: The IPU and Association of Secretaries General of Parliaments hold a one-day conference in Geneva titled Evaluating Parliament: Objectives, Methods, Results, and Impact.
- November: The Parliamentary Studies Centre of Australia National University holds a workshop titled Benchmarking of Parliamentary Performance in Canberra for the New Zealand parliament and the Australian national and state parliaments. The CPA holds a meeting on Pacific regional benchmarks during the Forum Presiding Officers and Clerks Annual Meeting in the Cook Islands. Preparation for the meeting includes national benchmarking exercises in Kiribati, Nauru, Niue, and Tuvalu.

box continues next page

Box 1A.1 Process Events Timeline *(continued)*

2010

- January: The CPA organizes a regional workshop in Dhaka on benchmarks for democratic parliaments for its Asia, India, and Southeast Asia regions.
- March: The World Bank, UNDP, and partners hold the International Conference on Benchmarking and Self-Assessment for Democratic Legislatures in Paris. Also in Paris, DFID, UNDP, and the World Bank hold the Third Donor Coordination Meeting on Parliamentary Development.

Note: Since the World Bank and Griffith University workshop in Brisbane in 2008, the partners have promoted an ongoing research agenda to pilot the different frameworks at the country level in both established and new legislatures, large and small legislatures. This effort has led to publications about individual parliaments but also to more comparative research, such as LSE (2009), which was prepared for the World Bank.

Annex 1B: Key Actors and Assessment Frameworks

Several organizations have been involved in the development of assessment frameworks. An outline of the NDI standards is provided in box 1B.1. Box 1B.2 outlines the CPA benchmarks, and box 1B.3 provides an overview of the APF criteria. Table 1B.1 displays the IPU framework.

Box 1B.1 Overview of Categories Covered in the National Democratic Institute of International Affairs Standards Publication

Part I: Election and Status of Legislators

1. Election and Status of Legislators

    1.1 The Election of Legislators

    1.2 Candidate Eligibility

    1.3 Incompatibility of Office

    1.4 Immunity

    1.5 Remuneration and Benefits

    1.6 Resignation

Part II: Organization of the Legislature

2. Procedure

    2.1 Rules of Procedure

    2.2 Sessions

    2.3 Plenary Agenda

    2.4 Plenary Debate

    2.5 Plenary Voting

    2.6 Presiding Officers

box continues next page

Box 1B.1 Overview of Categories Covered in the National Democratic Institute of International Affairs Standards Publication *(continued)*

3. Committees
   3.1 Organization
   3.2 Powers
4. Political Parties, Party Groups, and Interest Caucuses
   4.1 Political Parties
   4.2 Party Groups
   4.3 Interest Caucuses
5. Parliamentary Staff
   5.1 Authority
   5.2 Hiring and Promotion
   5.3 Organization and Management
   5.4 Media Function

Part III: Functions of the Legislature
6. Legislative Function
   6.1 In General
   6.2 Legislative Procedure
   6.3 Financial and Budgetary Powers
   6.4 Delegation of Legislative Power
   6.5 Constitutional Amendments
7. Oversight Function
   7.1 In General
   7.2 Commissions of Inquiry
   7.3 Legislative Ombudsmen
   7.4 Public Accounts Committees or Audit Committees
   7.5 No Confidence and Impeachment
   7.6 Legislative-Judicial Relationship
8. Representational Function
   8.1 Representational Nature of the Legislature
   8.2 Constituent Relations
   8.3 International Representation

Part IV: Values of the Legislature
9. Accessibility
   9.1 Citizens and the Press
   9.2 Languages and Disabilities
10. Transparency and Integrity
   10.1 Transparency and Integrity
   10.2 Pressure Groups and Lobbyists
11. Public Consultation and Participation
   11.1 Citizen Participation

Source: NDI 2007, 76–81.

Box 1B.2 Overview of Categories Covered in the CPA Recommended Benchmarks for Democratic Legislatures

I. General

 1. General

     1.1 Elections

     1.2 Candidate Eligibility

     1.3 Incompatibility of Office

     1.4 Immunity

     1.5 Remuneration and Benefits

     1.6 Resignation

     1.7 Infrastructure

II. Organization of the Legislature

 2. Procedure and Sessions

     2.1 Rules of Procedure

     2.2 Presiding Officers

     2.3 Convening Sessions

     2.4 Agenda

     2.5 Debate

     2.6 Voting

     2.7 Records

 3. Committees

     3.1 Organization

     3.2 Powers

 4. Political Parties, Party Groups, and Cross-Party Groups

     4.1 Political Parties

     4.2 Party Groups

     4.3 Cross-Party Groups

 5. Parliamentary Staff

     5.1 General

     5.2 Recruitment

     5.3 Promotion

     5.4 Organization and Management

III. Functions of the Legislature

 6. Legislative Function

     6.1 General

     6.2 Legislative Procedure

     6.3 The Public and Legislation

 7. Oversight Function

     7.1 General

     7.2 Financial and Budget Oversight

     7.3 No Confidence and Impeachment

box continues next page

Box 1B.2 Overview of Categories Covered in the CPA Recommended Benchmarks for Democratic Legislatures *(continued)*

8. Representational Function
   8.1 Constituent Relations
   8.2 Parliamentary Networking and Diplomacy

IV. Values of the Legislature
   9. Accessibility
      9.1 Citizens and the Press
      9.2 Languages
   10. Ethical Governance
      10.1 Transparency and Integrity

Source: CPA 2006.

Box 1B.3 Overview of Categories Covered in the APF Criteria

1. Elections and Status of Parliamentarians
   1.1 Elections
   1.2 Eligibility
   1.3 Status of Parliamentarians
   1.4 Material Situation of Parliamentarians
2. Rights and Duties of Parliament
   2.1 Organization of Parliamentary Work
   2.2 Legislative Function
   2.3 Parliamentary Oversight
   2.4 Parliamentary Committees
   2.5 International Relations
3. Organization of Parliament
   3.1 Status of Political Parties, Parliamentary Groups, and the Opposition
   3.2 Status of Administrative Staff
   3.3 Budget
   3.4 Material Resources
4. Parliament and Communications
   4.1 Accessibility of Parliament
   4.2 Dissemination of Parliamentary Information

Source: APF 2009.

Table 1B.1 IPU Framework on the Parliamentary Contribution to Democracy

Basic objectives or values	Requirements	Possible procedural and institutional means for the realization of these objectives or values
A parliament that is representative	An elected parliament that is socially and politically representative and committed to equal opportunities for its members so that they can carry out their mandates	• Free and fair electoral system and process, including means of ensuring representation of or by all sectors of society with a view to reflecting national and gender diversity, for example, by using special procedures to ensure representation of marginalized or excluded groups • Open, democratic, and independent party procedures, organizations, and systems • Mechanisms to ensure the rights of the political opposition and other political groups and to allow all members to exercise their mandates freely and without being subjected to undue influence and pressure • Freedom of speech and association, including guarantees of parliamentary rights and immunities, including the integrity of the presiding officers and other office holders • Equal opportunities policies and procedures, including nondiscriminatory hours and conditions of work and language facilities for all members
A parliament that is transparent	A parliament that is open to the nation and is transparent in the conduct of its business	• Proceedings open to the public • Prior information disseminated to the public on the business before parliament • Documentation available in relevant languages • Availability of user-friendly tools such as the World Wide Web • Public relations officers and facilities that belong to the parliament • Legislation on freedom of and access to information
A parliament that is accessible	Involvement of the public, including civil society and other people's movements, in the work of the parliament	• Effective electoral sanction and monitoring processes • Reporting procedures to inform constituents • Standards and enforceable code of conduct • Adequate salaries for members • Registration of outside interests and income • Enforceable limits on and transparency in election fundraising and expenditure
A parliament that is effective at all levels:	Effective organization of business in accordance with these democratic norms and values	• Mechanisms and resources to ensure the independence and autonomy of parliament, including parliament's control of its own budget • Availability of a nonpartisan professional staff separate from the main civil service

table continues next page

Table 1B.1 **IPU Framework on the Parliamentary Contribution to Democracy** (continued)

Basic objectives or values	Requirements	Possible procedural and institutional means for the realization of these objectives or values
		• Adequate unbiased research and information facilities for members, including parliament's own business committee, procedures for effective planning and timetabling of business, systems for monitoring parliamentary performance, and opinion surveys among relevant groups on perceptions of performance
National level	Effective performance of legislative and scrutiny functions and effective performance as a national forum for issues of common concern	• Systematic procedures for executive accountability, including adequate powers and resources for committees and accountability to parliament of nongovernmental public bodies and commissions • Mechanisms to ensure effective parliamentary engagement in the national budget process in all its stages, including the subsequent auditing of accounts • Ability to address issues of major concern to society, to mediate in the event of tension and prevent violent conflict, and to shape public institutions that cater for the needs of the entire population • For parliaments that approve senior appointments or perform judicial functions, mechanisms to ensure a fair, equitable, and nonpartisan process
International level	Active involvement of parliament in international affairs	• Procedures for parliamentary monitoring of and input into international negotiations as well as oversight of the positions adopted by the government • Mechanisms that allow for parliamentary scrutiny of activities of international organizations and input into their deliberations • Mechanisms for ensuring national compliance with international norms and the rule of law • Interparliamentary cooperation and parliamentary diplomacy
Local level	Cooperative relationships with state, provincial, and local legislatures	• Mechanisms for regular consultations between the presiding officers of the national and subnational parliaments or legislatures on national policy issues to ensure that decisions are informed by local needs

Source: IPU 2008, 25–27.

Annex 1C: International IDEA's Questions on the Democratic Effectiveness of Parliament

Overarching question: Does the parliament or legislature contribute effectively to the democratic process?

2.4.1 How independent is the parliament or legislature of the executive, and how freely are its members able to express their opinions?

2.4.2 How extensive and effective are the powers of the legislature to initiate, scrutinize, and amend legislation?

2.4.3 How extensive and effective are the powers of the legislature to oversee the executive and hold it to account?

2.4.4 How rigorous are the procedures for approval and supervision of taxation and public expenditure?

2.4.5 How freely are all parties and groups able to organize within the parliament or legislature and contribute to its work?

2.4.6 How extensive are the procedures of the parliament or legislature for consulting the public and relevant interests across the range of its work?

2.4.7 How accessible are elected representatives to their constituents?

2.4.8 How well does the parliament or legislature provide a forum for deliberation and debate on issues of public concern?

Annex 1D: The Frameworks: Commonalities and Differences

Table 1D.1 offers a comparison of three different frameworks: the CPA benchmarks, the NDI standards, and the APF criteria. The table uses a color-coded system. Benchmarks, standards, or criteria that match are coded as light gray, those that are very similar are coded as medium gray, and those that are new or that only appear in one set of standards are coded as dark gray.

Table 1D.1 Comparative Table of the CPA Benchmarks, NDI Standards, and APF Criteria

CPA benchmarks	NDI standards	APF criteria
1.1.1 Members of the popularly elected or only house shall be elected by direct universal and equal suffrage in a free and secret ballot.	1.1.1 Members of the popularly elected or only house shall be directly elected through universal and equal suffrage in a free and secret ballot.	1.1.1 The National Constitution shall include basic rules regarding elections and the status of legislators.
1.1.2 Legislative elections shall meet international standards for genuine and transparent elections.	1.1.2 Legislative elections shall meet international standards for genuine and transparent elections.	1.1.2 Legislators shall be elected by universal suffrage through an electoral process that is free, reliable, transparent, and in accordance with international and national standards. However, in a bicameral legislature, the second chamber may be governed by special regulations stipulated in the constitution or the laws of each country.
1.1.3 Term lengths for members of the popular house shall reflect the need for accountability through regular and periodic legislative elections.	1.1.3 Term lengths for members of the popular house shall reflect the need for accountability through regular and periodic legislative elections.	See criterion 1.1.2.
		1.1.3 Elections shall be held at regular intervals. Term limits shall be established for legislatures, and at the end of the term, new elections shall be held.
		1.1.4 Elections shall take place without any restriction or violation of freedom and security of person, freedom of opinion and speech, freedom of assembly and demonstration, and freedom of association of all voters and electoral candidates.
		1.1.5 The organization and management of elections, from the preparatory procedures and electoral campaign to vote tallying and the announcement of results, shall be the responsibility of bodies vested with the authority to closely monitor the electoral process, to ensure the fairness of the elections and the full participation of citizens in them, and to ensure the equal treatment of candidates throughout the electoral procedures.
		1.1.6 All legally constituted political parties shall have the right to participate in all stages of the electoral process, in accordance with the democratic principles set forth in the basic legal provisions and adhered to by institutions.
		1.1.7 An independent and impartial jurisdictional authority shall be responsible for management of electoral disputes.

table continues next page

Table 1D.1 Comparative Table of the CPA Benchmarks, NDI Standards, and APF Criteria *(continued)*

CPA benchmarks	NDI standards	APF criteria
1.2.1 Restrictions on candidate eligibility shall not be based on religion, gender, ethnicity, race, or disability.	1.2.1 Restrictions on candidate eligibility shall not be based on religion, gender, ethnicity, race, or physical ability.	1.2.1 Restrictions on candidate eligibility shall not be based on gender, race, language, religion, economic status, physical disability, or private life considerations.
1.2.2 Special measures to encourage the political participation of marginalized groups shall be narrowly drawn to accomplish precisely defined, and time-limited objectives.	1.2.2 Measures of positive discrimination used to encourage the political participation of marginalized groups shall be narrowly drawn to accomplish precisely defined and limited objectives.	1.2.2 Notwithstanding the preceding clause, special measures may be taken to ensure the representation of national or regional diversity and its components.
1.3.1 No elected member shall be required to take a religious oath against his or her conscience in order to take his or her seat in the legislature.	1.2.3 No elected member shall be required to take a religious oath against his/her conscience in order to take his/her seat in the legislature.	1.3.1.1 No elected member shall be required to take a religious oath against his or her conscience in order to take his or her seat in the legislature.
1.3.2 In a bicameral legislature, a legislator may not be a member of both houses.	1.3.1 In a bicameral legislature, a legislator may not be a member of both houses.	1.3.1.2 In a bicameral legislature, a legislator may not be a member of both chambers simultaneously.
		1.3.1.3 Incompatible parliamentary offices must be defined by law.
		1.3.1.4 A special procedure shall be established to monitor and sanction penalties.
1.3.3 A legislator may not simultaneously serve in the judicial branch or as a civil servant of the executive branch.	1.3.2 A legislator may not simultaneously serve in the judicial branch or as a civil servant of the executive branch, except in limited instances involving front-line delivery of public services.	
1.4.1 Legislators shall have immunity for anything said in the course of the proceedings of legislature.	1.4.1 Legislators shall have immunity for speech conducted during the exercise of their duties; former legislators shall never be liable for speech conducted during the exercise of their duties as a legislator.	1.3.2.2 No legislator shall be prosecuted, investigated, arrested, detained, or tried or imprisoned as a result of opinions expressed, orally or in writing, or votes cast in the performance of his or her duties.
1.4.2 Parliamentary immunity shall not extend beyond the term of office, but a former legislator shall continue to enjoy protection for his or her term of office.	1.4.2 Parliamentary immunity shall not be used to place legislators above the law and shall not extend beyond their term of office, though a former legislator shall continue to enjoy protection for his/her term of office.	1.3.2.3 Parliamentary immunity shall not extend beyond the term of office.
1.4.3 The executive branch shall have no right or power to lift the immunity of a legislator.	1.4.3 Only an act or vote of the legislature can lift parliamentary privilege and the immunity of a legislator. The executive branch shall have no right or power to lift the immunity of a legislator.	1.3.2.4 The decision to lift the immunity of a legislator is the sole purview of the legislature.

table continues next page

Table 1D.1 Comparative Table of the CPA Benchmarks, NDI Standards, and APF Criteria *(continued)*

CPA benchmarks	NDI standards	APF criteria
	1.4.4 After the legislature votes to lift the immunity of a legislator, it has no power to mandate changes to or otherwise affect proceedings involving the legislator before other branches of government.	
1.4.4 Legislators must be able to carry out their legislative and constitutional functions in accordance with the constitution, free from interference.		1.3.2.1 Legislators must be able to perform their duties free from any undue influence or pressure.
1.5.1 The legislature shall provide proper remuneration and reimbursement of parliamentary expenses to legislators for their service, and all forms of compensation shall be allocated on a nonpartisan basis.	1.5.1 The legislature shall provide all legislators with fair remuneration and adequate physical infrastructure, and all forms of remuneration and infrastructure shall be allocated on a nonpartisan basis.	1.4.1.1 The legislature shall provide all legislators with fair remuneration and proper material infrastructure to facilitate fulfillment of their mandates as well as reimbursement for expenses incurred in the performance of their duties.
		1.4.1.2 All forms of remuneration paid to legislators by the legislature shall be allocated in a transparent manner on the basis of the duties performed.
1.6.1 Legislators shall have the right to resign their seats.	1.6.1 Legislators shall have the right to resign their positions.	
1.7.1 The legislature shall have adequate physical infrastructure to enable members and staff to fulfill their responsibilities.		3.4.1.1 The legislature shall have access to adequate physical and material infrastructure to enable its members to fulfill their mandates in satisfactory conditions.
2.1.1 Only the legislature may adopt and amend its rules of procedure.	2.1.1 Only the legislature may adopt and amend its rules of procedure.	2.1.1.1 Every legislature—or, as the case may be, each of the chambers of the legislature—shall draft, adopt, and amend its rules of procedure.
		2.1.1.2 Legislative rules of procedure—or, as the case may be, the rules of procedure of each of the chambers of the legislature—shall be consistent with the constitution.
		2.1.1.3 The legislature shall take significant measures to establish and preserve gender balance in its various bodies at all levels of responsibility.

table continues next page

Table 1D.1 Comparative Table of the CPA Benchmarks, NDI Standards, and APF Criteria (continued)

CPA benchmarks	NDI standards	APF criteria
2.2.1 The legislature shall select or elect presiding officers pursuant to criteria and procedures clearly defined in the rules of procedure.	2.6.1 The legislature shall elect or select presiding officers and members of a steering body pursuant to criteria and procedures clearly defined in the rules of procedure.	2.1.2.1 The legislature—or, as the case may be, each of the chambers of the legislature—shall elect a chair and at least one vice chair pursuant to the procedures defined in its rules of procedure.
2.3.1 The legislature shall meet regularly, at intervals sufficient to fulfill its responsibilities.	2.2.1 The legislature shall meet regularly, at intervals sufficient to fulfill its responsibilities.	2.1.3.1 Legislative sessions shall be held at sufficiently regular intervals to allow the legislature to properly fulfill its responsibilities.
2.3.2 The legislature shall have procedures for calling itself into regular session.		See criterion 2.1.3.2
2.3.3 The legislature shall have procedures for calling itself into extraordinary or special session.	2.2.2 The legislature shall have and follow procedures for calling itself into extraordinary or special session.	2.1.3.2 The legislature shall draft rules of procedures for calling itself into regular or extraordinary session.
2.3.4 Provisions for the executive branch to convene a special session of the legislature shall be clearly specified.	2.2.3 Provisions for the executive branch to convene a special session of the legislature shall be clearly specified.	2.1.3.5 Provisions for the executive branch or for a group of legislators to convene a special session of the legislature shall be clearly specified.
		2.1.4.1 Public sessions shall be scheduled in such a way as to allow adequate time for review of the items on the legislative agenda. 2.1.4.2 As far as possible, public sessions shall avoid time conflicts with the meetings of other legislative bodies.
2.4.1 Legislators shall have the right to vote to amend the proposed agenda for debate.	2.3.1 Legislators shall have the right to vote to amend the proposed agenda for debate.	2.1.5.1 Legislators shall have the right to set the agenda and the time allocated to each of the points under review. 2.1.5.2 Establishment of the agenda shall be entrusted to a legislative body.
2.4.2 Legislators in the lower or only house shall have the right to initiate legislation and to offer amendments to proposed legislation.	2.3.2 Legislators in the lower or popularly elected chamber shall have the right to initiate legislation and to offer amendments to proposed legislation.	2.1.5.6 Legislators in the lower or popularly elected chamber shall have the right to introduce legislation and amendments.
2.4.3 The legislature shall give legislators adequate advance notice of session meetings and the agenda for the meeting.	2.3.3 The legislature shall give legislators and citizens adequate advance notice of session meetings and the agenda for the meeting.	2.1.5.3 The legislature shall give legislators adequate advance notice of session meetings and the agenda for the meetings. 2.1.5.4 A timetable for legislative work shall be established so that the legislative schedule is known.

table continues next page

Table 1D.1 Comparative Table of the CPA Benchmarks, NDI Standards, and APF Criteria (continued)

CPA benchmarks	NDI standards	APF criteria
2.5.1 The legislature shall establish and follow clear procedures for structuring debate and determining the order of precedence of motions tabled by members.	2.4.1 The legislature shall create and follow clear procedures for structuring debate and determining the order of precedence of motions tabled by members.	2.1.5.5 The agenda shall facilitate the review of draft legislation and proposals within a reasonable timeframe and allow legislators to engage in meaningful debate of such legislation. 2.2.5.1 The legislature shall establish and follow clear procedures for structuring legislative debates and determining the order of precedence of motions tabled by members.
2.5.2 The legislature shall provide adequate opportunity for legislators to debate bills prior to a vote.	2.4.2 The legislature shall provide meaningful opportunity for legislators to publicly debate bills prior to a vote.	2.2.5.2 The legislature shall provide adequate opportunity for legislators to publicly debate draft legislation and proposals prior to a vote.
2.6.1 Plenary votes in the legislature shall be public.	2.5.1 There shall be a presumption that votes in the legislature shall be public; the legislature shall publicly codify any exceptions to the presumption and give advance notice before a nonpublic vote.	2.2.6.1 Unless explicitly stated otherwise, plenary votes in the legislature shall be public. 2.2.7.4 Debate on draft legislation and proposals shall be open to the public. 4.1.2.2 Plenary sessions in the legislature shall be public.
2.6.2 Members in a minority on a vote shall be able to demand a recorded vote.	2.5.2 The legislature shall establish and follow procedures for a minority of legislators to demand that a recorded method of voting be used.	
2.6.3 Only legislators may vote on issues before the legislature.	2.5.3 Only legislators shall have a vote on issues before the legislature.	2.2.6.2 Only legislators may vote on issues before the legislature. 2.2.6.3 Voting shall be private and not mandatory. 2.2.6.4 Unless otherwise clearly stipulated by law, delegation of voting rights shall be forbidden.
2.7.1 The legislature shall maintain and publish readily accessible records of its proceedings.	5.4.2 The legislature shall maintain a central depository for records of daily proceedings and votes that can be readily accessed by legislators, staff, and citizens.	2.2.7.3 Information on legislation shall be readily available to both legislators and citizens.
3.1.1 The legislature shall have the right to form permanent and temporary committees.	3.1.1 The legislature shall have the right to form permanent and temporary committees.	2.4.1.1 The legislature's rules of procedure shall provide for the option to form permanent or temporary committees. 2.4.1.3 Legislative work and voting procedures shall be consistent with the legislature's rules of procedure.

table continues next page

Table 1D.1 Comparative Table of the CPA Benchmarks, NDI Standards, and APF Criteria (continued)

CPA benchmarks	NDI standards	APF criteria
		2.4.1.4 The legislature's rules of procedure shall specify the mandate and composition of committees.
		2.4.1.5 The areas of competence of committees shall be clearly defined to avoid conflicts of jurisdiction.
3.1.2 The legislature's assignment of committee members on each committee shall include both majority and minority party members and reflect the political composition of the legislature.	3.1.2 The legislature's assignment of committee seats shall reflect the political party composition of the legislature and shall include both majority and minority party members.	2.4.2.1 The composition of committees shall reflect as closely as possible the composition of the legislature, particularly with respect to gender.
		2.4.2.2 A committee shall choose or elect a chair and at least one vice chair in accordance with the procedure defined in the legislature's rules of procedure.
3.1.3 The legislature shall establish and follow a transparent method for selecting or electing the chairs of committees.	3.1.3 The legislature shall establish and follow a transparent method for electing or selecting the chairs of committees.	2.4.1.2 Committee hearings shall be public when so stipulated in the legislature's rules of procedure. Any exceptions to this rule shall be clearly defined and provided for in the rules of procedure.
3.1.4 Committee hearings shall be in public. Any exceptions shall be clearly defined and provided for in the rules of procedure.	3.1.4 There shall be a presumption that committee hearings are open to the general public; the legislature shall publicly codify any exceptions to the presumption and give advance notice before a nonpublic committee meeting.	2.4.1.6 The legislature's rules of procedure shall set forth the conditions under which committees can hold public hearings.
3.1.5 Votes of committee shall be in public. Any exceptions shall be clearly defined and provided for in the rules of procedure.		
3.2.1 There shall be a presumption that the legislature will refer legislation to a committee, and any exceptions must be transparent, narrowly defined, and extraordinary in nature.	3.2.1 There shall be a presumption that the legislature will refer legislation to a committee, and any exceptions must be transparent, narrowly defined, and extraordinary in nature.	2.4.3.1 The legislature shall refer the review of draft legislation and proposals to a committee. Any exception to this rule shall be stipulated in the legislature's rules of procedure.
3.2.2 Committees shall scrutinize legislation referred to them and have the power to recommend amendments or amend the legislation.	3.2.2 All committees shall have the power to amend legislation.	2.4.3.2 Committees shall review the draft legislation and proposals submitted to them and have the authority to make amendments thereto.
3.2.3 Committees shall have the right to consult and/or employ experts.	3.2.3 All committees shall have the right to consult and/or hire experts.	2.4.2.3 All committees shall have the right to hire experts.

table continues next page

Table 1D.1 Comparative Table of the CPA Benchmarks, NDI Standards, and APF Criteria (continued)

CPA benchmarks	NDI standards	APF criteria
3.2.4 Committees shall have the power to summon persons, papers, and records, and this power shall extend to witnesses and evidence from the executive branch, including officials.	3.2.4 Committees shall have the power of summons to examine persons, papers, and records, including witnesses and evidence from the executive branch.	2.4.3.3 Committees shall have the power to summon witnesses and documents they require to carry out their work.
3.2.5 Only legislators appointed to the committee, or authorized substitutes, shall have the right to vote in committee.	3.2.5 Only legislators appointed to the committee shall have the right to vote in the committee.	2.4.3.4 Only legislators appointed to a committee shall have the right to vote in the committee.
3.2.6 Legislation shall protect informants and witnesses presenting relevant information to commissions of inquiry about corruption or unlawful activity.	7.1.4 "Whistleblower" protections shall protect informants and witnesses presenting accurate information about corruption or unlawful activity.	2.4.2.4 Witnesses summoned by the committees of inquiry shall have the right to protection.
4.1.1 The right of freedom of association shall exist for legislators, as for all people.	4.1.1 The right of freedom of association shall exist for legislators, as for all people.	
4.1.2 Any restrictions on the legality of political parties shall be narrowly drawn in law and shall be consistent with the International Covenant on Civil and Political Rights.	4.1.2 Any restrictions on the legality of political parties shall be narrowly drawn in law and shall be consistent with the International Covenant on Civil and Political Rights.	3.1.1.1 Where it exists, public and private financing of political parties shall comply with the norms of transparency. A competent and independent judicial authority shall supervise such financing. Equal access to public financing shall be ensured.
4.2.1 Criteria for the formation of parliamentary party groups, and their rights and responsibilities in the legislature, shall be clearly stated in the rules.	4.2.1 Criteria for the formation of parliamentary party groups, and their rights and responsibilities in the legislature, shall be clearly stated in the rules.	3.1.2.2 Criteria for the formation of a parliamentary party group, as well as the group's rights and responsibilities in the legislature, shall be clearly stated in the rules of procedure.
	4.2.2 In a non-party-list electoral system, membership of a parliamentary party group shall be voluntary and a legislator shall not lose his/her seat for leaving his/her party group.	
		3.1.2.1 Parliamentary party groups shall have legal status or another form of recognition.

table continues next page

Table 1D.1 Comparative Table of the CPA Benchmarks, NDI Standards, and APF Criteria *(continued)*

CPA benchmarks	NDI standards	APF criteria
4.2.2 The legislature shall provide adequate resources and facilities for party groups pursuant to a clear and transparent formula that does not unduly advantage the majority party.	4.2.3 The legislature shall provide adequate resources and facilities for party groups pursuant to a clear and transparent formula that does not unduly advantage the majority party.	3.1.2.3 All parliamentary party groups shall have the right to place items on the agenda, to participate in debates, and to propose amendments to draft laws.
		3.1.2.4 The legislature shall provide adequate resources and facilities to parliamentary party groups in an equitable manner.
4.3.1 Legislators shall have the right to form interest caucuses around issues of common concern.	4.3.1 Legislators shall have the right to form interest caucuses around issues of common concern.	
5.1.1 The legislature shall have an adequate nonpartisan professional staff to support its operations including the operations of its committees.		3.2.1.1 Administrative management of the legislature shall be carried out by a permanent, professional, and nonpartisan staff to support the operations of its committees.
5.1.2 The legislature, rather than the executive branch, shall control the parliamentary service and determine the terms of employment.	5.1.1 The legislature, rather than the executive branch, shall control its staff.	3.2.1.2 The legislature shall have control of parliamentary services and shall determine the terms of employment of its staff, independently from the executive branch.
		3.2.1.3 The legislature shall performs its duties with impartiality and be mindful of its duty of restraint.
		3.2.1.5 Women shall be represented at all levels of the legislature's administration.
5.1.3 The legislature shall draw and maintain a clear distinction between partisan and nonpartisan staff.	5.1.2 The legislature shall draw and maintain a clear distinction between partisan and nonpartisan staff.	3.2.1.4 A clear distinction shall be drawn between partisan and nonpartisan staffs (staff working exclusively for a legislator or a political group and employed by the legislator).
5.1.4 Members and staff of the legislature shall have access to sufficient research, library, and ICT [information and communication technology] facilities.		

table continues next page

Table 1D.1 Comparative Table of the CPA Benchmarks, NDI Standards, and APF Criteria *(continued)*

CPA benchmarks		NDI standards		APF criteria	
5.2.1	The legislature shall have adequate resources to recruit staff sufficient to fulfill its responsibilities. The rates of pay shall be broadly comparable to those in the public service.	5.2.1	The legislature shall have adequate resources to hire a staff sufficient to fulfill its responsibilities. Nonpartisan staff shall be recruited and promoted on the basis of merit and equal opportunity.	3.2.2.1	The legislature shall have adequate resources to hire a staff sufficient to fulfill its responsibilities.
				3.2.2.2	The salary scale for parliamentary staff shall be comparable to that of the public service.
5.2.2	The legislature shall not discriminate in its recruitment of staff on the basis of race, ethnicity, religion, gender, disability, or, in the case of nonpartisan staff, party affiliation.	5.2.2	The legislature shall not discriminate in its hiring of any staff on the basis of race, ethnicity, religion, gender, or physical ability. Additionally, it shall not discriminate in its hiring of nonpartisan staff on the basis of party affiliation.		
5.3.1	Recruitment and promotion of nonpartisan staff shall be on the basis of merit and equal opportunity.		See standard 5.2.1	3.2.2.3	The recruitment and promotion of parliamentary staff shall be based on a fair and transparent process.
5.4.1	The head of the parliamentary service shall have a form of protected status to prevent undue political pressure.	5.3.1	The legislature shall clearly codify the responsibilities of the semi-independent, nonpartisan secretary general. The secretary general shall be ultimately accountable to the legislature, and the secretary general's tenure shall outlast the legislature.	3.2.3.1	The status of parliamentary staff members shall protect them from any form of undue political pressure.
5.4.2	Legislatures should, either by legislation or resolution, establish corporate bodies responsible for providing services and funding entitlements for parliamentary purposes and providing for governance of the parliamentary service.	5.3.2	No partisan or nonpartisan staff of the legislature, including the secretary general, shall have any legislative or procedural authority, including voting, in the legislature.		
5.4.3	All staff shall be subject to a code of conduct.	5.3.3	All staff shall be subject to a code of conduct.	3.2.3.2	A mechanism shall be put in place to deter, detect, and bring to justice all partisan and nonpartisan staff engaging in fraudulent or corrupt practices.

table continues next page

Table 1D.1 Comparative Table of the CPA Benchmarks, NDI Standards, and APF Criteria *(continued)*

CPA benchmarks	NDI standards	APF criteria
	5.4.3 Nonpartisan staff shall publish transcripts, votes, and schedules.	
6.1.1 The approval of the legislature is required for the passage of all legislation, including budgets.	6.1.1 The approval of the legislature is required for the passage of all legislation, including budgets.	2.2.1.1 The legislature shall vote on all legislation, including the budget. Any exception to this rule shall be explicitly stated.
	6.3.1 The proposed national budget shall require the approval of the legislature, and the legislature shall have the power to amend the budget before approving it.	2.2.2.1 The legislature shall adopt a clearly established legislative procedure for the submission of legislation, its review by the legislature, and its promulgation.
6.1.2 Only the legislature shall be empowered to determine and approve the budget of the legislature.	6.3.3 Only the legislature shall be empowered to determine and approve the budget of the legislature.	3.3.1.1 Only the legislature shall be empowered to determine and approve its own budget, and the executive branch shall not determine the resources needed by the legislature to fulfill its responsibilities.
6.1.3 The legislature shall have the power to enact resolutions or other nonbinding expressions of its will.	6.1.2 The legislature shall have the power to enact resolutions or other nonbinding expressions of its will.	2.2.1.2 The legislature shall have the power to enact resolutions without notice and take a position on matters of general interest.
	6.4.1 The legislature shall have the prerogative to delegate legislative functions to the executive branch under legally grounded criteria, for a limited period of time, and for strictly defined purposes.	
6.1.4 In bicameral systems, only a popularly elected house shall have the power to bring down government.	7.5.2 Chambers where a majority of members are not directly elected shall have no power or means to collapse the government.	
6.1.5 A chamber where a majority of members are not directly or indirectly elected may not indefinitely deny or reject a money bill.	6.1.3 Chambers where a majority of members are appointed and/or enjoy hereditary seats shall have no power or means to permanently deny or reject money bills.	
6.2.1 In a bicameral legislature, there shall be clearly defined roles for each chamber in the passage of legislation.	6.2.1 In a bicameral legislature, the legislature shall clearly define the roles of each chamber in the passage of legislation.	2.2.2.2 In a bicameral legislature, there shall be clearly defined roles for each chamber.

table continues next page

Table 1D.1 Comparative Table of the CPA Benchmarks, NDI Standards, and APF Criteria (continued)

CPA benchmarks	NDI standards	APF criteria
		2.2.2.3 In a bicameral legislature, a conciliation process must be in place to resolve potential disagreements between the two chambers.
		2.2.3.1 An independent judiciary shall be made responsible for constitutional review—that is, for verifying if laws that have been enacted are consistent with the constitution.
		2.2.4.1 All legislators have the right to propose amendments, in accordance with the rules governing their admissibility.
		2.2.4.2 Specific regulatory provisions shall stipulate the order of amendments and the procedure for discussing them in order to facilitate organized debate and allow all opinions to be expressed.
6.2.2 The legislature shall have the right to override an executive veto.	6.2.2 The legislature shall have the right to override an executive veto.	
6.3.1 Opportunities shall be given for public input into the legislative process.	11.1.1 The legislature shall create and utilize mechanisms for receiving and considering public views on proposed legislation.	2.2.7.1 Citizens shall have input in the legislative process, in particular through their parliamentary representative.
6.3.2 Information shall be provided to the public in a timely manner regarding matters under consideration by the legislature.	11.1.2 Information shall be provided to the public in a timely manner regarding matters under consideration by the legislature, sufficient to allow the public and civil society to provide their views on draft legislation.	2.2.7.2 Citizens shall be informed in a timely manner of issues being debated by the legislature.
	6.5.1 In the absence of a public referendum, constitutional amendments shall require the approval of the legislature.	
	7.1.1 The legislature shall have sufficient means and mechanisms to effectively fulfill its oversight function.	2.3.1.1 The legislature shall be empowered to oversee the actions of the government.
7.1.1 The legislature shall have mechanisms to obtain information from the executive branch sufficient to exercise its oversight function in a meaningful way.	7.1.2 The legislature shall have mechanisms to obtain information from the executive branch sufficient to meaningfully exercise its oversight function.	2.3.1.2 The government shall provide the legislature with sufficient access to the information necessary to effectively exercise its oversight function.

table continues next page

Table 1D.1 Comparative Table of the CPA Benchmarks, NDI Standards, and APF Criteria (continued)

CPA benchmarks	NDI standards	APF criteria
		2.3.1.3 A rigorous, systematic procedure shall be established to govern written and oral questions addressed to the executive branch by legislators.
7.1.2 The oversight authority of the legislature shall include meaningful oversight of the military security and intelligence services.	7.1.3 The oversight authority of the legislature shall include meaningful oversight of the security and intelligence forces and of state-owned enterprises.	2.3.1.4 In addition to supervising ministries, the legislature shall oversee state-owned enterprises and government agencies, including those in the defense and national security sectors.
7.1.3 The oversight authority of the legislature shall include meaningful oversight of state-owned enterprises.		
7.2.1 The legislature shall have a reasonable period of time in which to review the proposed national budget.	6.3.2 The legislature shall have a reasonable period of time in which to review the proposed budget.	2.3.2.1 The legislature shall have a reasonable period of time in which to review and discuss the national budget.
7.2.2 Oversight committees shall provide meaningful opportunities for minority or opposition parties to engage in effective oversight of government expenditures. Typically, the public accounts committee will be chaired by a member of the opposition party.	7.4.1 The legislature shall ensure that public accounts committees provide opposition parties with a meaningful opportunity to engage in effective oversight of executive branch expenditures.	2.3.2.2 Parliamentary committees shall allow for effective oversight of government expenditures by all parliamentary party groups in accordance with the legislature's rules of procedure.
	7.2.1 The law shall guarantee the right of the legislature to create commissions of inquiry. Such commissions shall have the power to compel executive branch officials to appear and give evidence under oath.	
7.2.3 Oversight committees shall have access to records of executive branch accounts and related documentation sufficient to be able to meaningfully review the accuracy of executive branch reporting on its revenues and expenditures.	7.4.2 Public accounts or audit committees shall have access to records of executive branch accounts and related documentation sufficient to be able to meaningfully review the accuracy of executive branch reporting on its revenues and expenditures.	2.3.2.3 The parliamentary committees specifically tasked with reviewing government expenditures shall have access to all of the documentation necessary as well as the power to hear high-ranking officials from the ministries and government agencies in order to conduct effective oversight of the expenditures of the executive branch.
7.2.4 There shall be an independent, nonpartisan supreme or national audit office whose reports are tabled in the legislature in a timely manner.	7.4.3 There shall be an independent, nonpartisan supreme or national audit office that conducts audits and reports to the legislature in a timely way.	2.3.2.4 An independent, nonpartisan body (court of auditors or auditor general) shall be established and provided with adequate resources and the authority required to carry out supervisory, audit, and oversight functions.

table continues next page

Table 1D.1 Comparative Table of the CPA Benchmarks, NDI Standards, and APF Criteria *(continued)*

CPA benchmarks	NDI standards	APF criteria
7.2.5 The supreme or national audit office shall be provided with adequate resources and legal authority to conduct audits in a timely manner.		2.3.2.5 The legislature shall receive reports from this body within a reasonable timeframe to allow for effective follow-up. See criterion 2.3.2.4.
	7.3.1 The legislature shall have a nonpartisan ombudsman or a similar body that investigates complaints of executive branch malfeasance, makes recommendations, and reports directly to the legislature.	2.3.2.6 The legislature shall have the right to seek the assistance of this body.
	7.5.1 The legislature shall have mechanisms to impeach or censure officials of the executive branch and/or express no confidence in the government.	2.3.3.1 Institutions shall ensure that clearly defined mechanisms are in place to establish balance between the legislative and executive branches.
7.3.1 The legislature shall have mechanisms to impeach or censure officials of the executive branch or express no confidence in the government.		
7.3.2 If the legislature expresses no confidence in the government, the government is obliged to offer its resignation. If the head of state agrees that no other alternative government can be formed, a general election should be held.	7.6.1 The legislature's consent shall be required in the confirmation of senior judges; and the legislature shall have mechanisms to impeach judges for serious crimes.	

table continues next page

Table 1D.1 Comparative Table of the CPA Benchmarks, NDI Standards, and APF Criteria (continued)

CPA benchmarks	NDI standards	APF criteria
8.1.1 The legislature shall provide all legislators with adequate and appropriate resources to enable the legislators to fulfill their constituency responsibilities.	8.1.1 The number of seats in the legislature shall not be so low, and hence the citizen-legislator ratio so high, as to render impossible meaningful constituent relations.	
	8.2.1 The legislature shall provide all legislators with sufficient resources to enable the legislators to fulfill their constituency responsibilities, including travel to and from their constituencies.	
8.2.1 The legislature shall have the right to receive development assistance to strengthen the institution of parliament.	8.3.1 The legislature, including its members and staff, shall have the right to send and receive development assistance, whether technical or advisory in nature, regardless of origin or destination.	2.5.3.1 Resources permitting, legislatures shall have the right to provide technical assistance to other legislatures.
8.2.2 Members and staff of parliament shall have the right to receive technical and advisory assistance, as well as to network and exchange experience with individuals from other legislatures.		2.5.3.2 Members and staff of parliament shall have the right to receive technical assistance.
		2.5.1.2 Legislators may participate in the activities of other entities and in events that offer the opportunity to share their experiences with members of other legislatures.
		2.5.1.1 In the context of parliamentary diplomacy, delegations shall reflect the composition of the legislature as closely as possible, in particular with respect to gender.
		2.5.1.3 Legislators shall have the right to participate in missions to other parliaments and to welcome foreign parliamentary delegations.
		2.5.1.4 The legislature shall fulfill its obligations to international parliamentary institutions.
		2.5.2.1 The legislature may participate in regional and international organizations to strengthen the legislative component of these organizations.

table continues next page

Table 1D.1 Comparative Table of the CPA Benchmarks, NDI Standards, and APF Criteria *(continued)*

CPA benchmarks	NDI standards	APF criteria
		2.5.2.2 The legislature shall have access to the information, organization, and resources necessary to review international issues.
		2.5.2.3 Legislators shall have the right to be a part of government delegations during missions or international negotiations.
9.1.1 The legislature shall be accessible and open to citizens and the media, subject only to demonstrable public safety and work requirements.	9.1.1 The legislature shall ensure that the buildings of the legislature shall be accessible and open to citizens and the press, subject only to demonstrable public safety and work requirements.	4.1.2.1 The legislature shall be accessible to the public, provided public security and the legislature's work are not jeopardized.
9.1.2 The legislature should ensure that the media are given appropriate access to the proceedings of the legislature without compromising the proper functioning of the legislature and its rules of procedure.		4.1.1.1 The legislature shall ensure that the media are given appropriate access to the public proceedings of the legislature without compromising the proper functioning of the legislature.
	9.1.2 The legislature shall not use credentialing of the media in the legislature for the purpose or with the effect of creating a ruling party bias.	4.1.1.2 Access by the media to the legislature shall be based on nonpartisan and transparent criteria.
9.1.3 The legislature shall have a nonpartisan media relations facility.	5.4.1 The legislature shall have a nonpartisan media relations facility that shall be sufficiently and consistently funded under the administrative budget and operate under the office of the secretary general.	
9.1.4 The legislature shall promote the public's understanding of the work of the legislature.		4.1.2.3 The legislature shall have access to resources to help citizens understand its proceedings.
		4.2.1.1 The legislature must foster a spirit of tolerance and promote all aspects of democratic culture to educate and raise awareness among public officials, all political actors, and all citizens about the ethical requirements of democracy and human rights.
		4.2.2.1 Laws, draft legislation and proposals, committee reports, and all other parliamentary documentation provided for by the legislature's rules of procedure shall be made accessible to the public.

table continues next page

Table 1D.1 Comparative Table of the CPA Benchmarks, NDI Standards, and APF Criteria (continued)

CPA benchmarks	NDI standards	APF criteria
9.2.1 Where the constitution or parliamentary rules provide for the use of multiple working languages, the legislature shall make every reasonable effort to provide for simultaneous interpretation of debates and translation of records.	9.2.1 The legislature shall facilitate the use of all working languages recognized by the constitution or in the rules of procedure, including simultaneous interpretation in debates and proceedings and the enactment of laws in all working languages.	4.2.3.1 The legislature shall promote awareness of its work through publicly available communication and information tools.
		4.1.3.1 Where the constitution or the legislature's rules of procedure provide for the use of multiple working languages, the legislature shall make every reasonable effort to ensure mutual understanding by all members of parliament.
	9.2.2 The legislature shall make every reasonable effort to publish all official papers and bills in all working languages recognized by the constitution or in the rules of procedure.	
	9.2.3 The legislature shall make every reasonable effort to accommodate the special needs of persons with disabilities, including wheelchair access, the translation of documents into Braille, and the use of closed captioning in televised broadcasts.	
10.1.1 Legislators should maintain high standards of accountability, transparency, and responsibility in the conduct of all public and parliamentary matters.		
10.1.2 The legislature shall approve and enforce a code of conduct, including rules on conflicts of interest and the acceptance of gifts.	10.1.3 To protect the dignity of the legislature, the legislature shall promulgate and enforce rules to regulate the conduct of legislators.	1.4.2.1 Where they are not specified by the constitution or by law, the legislature may establish rules governing the transparency and conduct of public and parliamentary activities, by which every legislator shall be bound.
	10.2.1 The legislature shall create a system for recording and making public all activities with, and exchange of gifts or favors between, lobbyists and legislators/legislative staff.	1.4.2.5 The legislature shall create a legal mechanism to govern relations between legislators or legislative staff and interest groups. This mechanism may be a public register of these interest groups and their activities.
	10.1.1 The legislature shall approve and enforce rules on conflicts of interest that promote the independence of legislators from private interests or unreasonable political pressures.	

table continues next page

Table 1D.1 Comparative Table of the CPA Benchmarks, NDI Standards, and APF Criteria *(continued)*

CPA benchmarks	NDI standards	APF criteria
10.1.3 Legislatures shall require legislators to fully and publicly disclose their financial assets and business interests.	10.1.2 Legislatures shall require legislators to fully disclose their financial assets and business interests.	1.4.2.2 Legislators shall avoid placing themselves in situations in which their personal interests may influence the performance of their duties.
		1.4.2.3 An asset declaration procedure has been established for legislators.
10.1.4 There shall be mechanisms to prevent, detect, and bring to justice legislators and staff engaged in corrupt practices.	10.1.4 The legislature shall create legal mechanisms to prevent, detect, and bring to justice legislators and staff engaged in corrupt practices.	1.4.2.4 The legislation shall include mechanisms to prevent and sanction corrupt practices by legislators.

Sources: APF 2009; CPA 2006; NDI 2007.

Annex 1E: Regional Perspectives

Regional benchmark discussions have affirmed existing benchmarks and standards while developing new and sometimes regionally specific benchmarks. This annex reviews regional trends in parliamentary assessment frameworks.

The Pacific

In 2008, the Forum Presiding Officers and Clerks Conference (FPOCC) mandated that its secretariat work with the CPA, the UNDP Pacific Centre, and other organizations on a Pacific version of benchmarks for democratic legislatures. Pacific legislators subsequently participated in a June 2009 workshop on benchmarks in Brisbane, Australia, and benchmark self-assessments were undertaken by the parliaments of Kiribati, Nauru, Niue, and Tuvalu. Finally, in cooperation with the CPA, the 2009 Conference of the Pacific Legislatures for Population and Governance (formerly the FPOCC) adopted the Pacific Islands Benchmarks for Democratic Legislatures. Among the major additions to the original CPA benchmarks (on which the Pacific Islands benchmarks are based) are the following:

- *1.5.2* An independent body should determine the appropriate remuneration, benefits, and other statutory entitlements of legislators.[32]
- *7.1.1* The legislature shall have appropriate legislation or a constitutional provision that clearly determines the size of cabinet, which should not exceed one-third of the total membership of the legislature.[33]
- *7.1.5* The oversight authority of the legislature shall include meaningful oversight of compliance with international human rights instruments and national constitutional rights, including consideration of gender and socioeconomic impact.
- *7.1.6* The oversight authority of the legislature shall include meaningful and timely oversight of accountability institutions, such as election commissions, human rights commissions, anticorruption commissions, ombudsmen, information commissions, and offices of auditors general.
- 7.2.2 The legislature shall have clear procedures requiring the government to provide timely responses to parliamentary committee reports and recommendations.

CPA Asian Regions (Asia, India, and Southeast Asia)

Members of parliament and regional secretaries and clerks from the CPA's Asia regions (Asia, India, and Southeast Asia) were hosted by the parliament of Bangladesh in Dhaka for the CPA Regional Workshop on Benchmarks for Democratic Parliaments, held on January 25–29, 2010. The regions established a process by which the regional secretaries and clerks undertook a first review of the original CPA benchmarks and recommended potential changes. In preparation for this activity, several of the clerks had examined their own parliaments' adherence to the benchmarks. The regional secretaries and clerks were then joined by members of parliament, who reviewed the original CPA benchmarks along with the clerks' recommendations and finalized the Recommended Benchmarks for Asia, India, and South-East Asia Regions' Democratic Legislatures (CPA 2010).

Participants affirmed the majority of the original benchmarks with some amendments. However, after much debate, they deleted CPA benchmarks 1.3.1 (on the grounds that two of the countries present constitutionally required members of parliament to take a religious oath); 3.1.4, 3.1.5, and 3.2.6 (on the grounds that holding committee hearings and votes in public was not common practice in the region); and 4.3.1[34] and 5.1.3 (which were considered redundant because the parliamentary staff was implicitly understood to be nonpartisan by the workshop participants). Benchmarks 7.1.2 and 7.1.3 were combined into a new benchmark 7.1.2, which states, "the oversight authority of the legislature shall include meaningful oversight of the security services and state owned enterprises," and a footnote was added defining *security services* as in the publication *Security System Reform and Governance* (OECD 2005). Although the workshop participants agreed to the principle of the independence of the parliamentary service, they recognized cases within the region where members of the parliamentary service were part of the broader public service. Hence, benchmark 5.1.2 was amended to read, "The legislature shall have an independent parliamentary service. In instances where parliamentary services are drawn from the public service, there shall be adequate safeguards to ensure noninterference from the Executive."

Finally, two new benchmarks were added:

- *1.1.4* Election expenses of candidates shall be monitored by the Election Commission or similar authority.
- *10.1.5* Legislatures should establish a mechanism to oversee the conduct of legislators.

Southern Africa

The SADC Parliamentary Forum's Draft Benchmarks for Democratic Legislatures in Southern Africa are still being discussed, as there is a need for further feedback from the lusophone members before they can be adopted.[35] Again, the SADC PF draft affirms many of the original CPA benchmarks, but it also includes scores of additional benchmarks, some of which are specific to regional issues (such as the use of constituency development funds or parliamentary approval of international loans). Other draft benchmarks attempt to address issues of concern to some parliaments in the region, such as floor crossing.

Like the APF, SADC PF has drafted additional benchmarks on elections and participation in international affairs. Many of the parliaments in the region receive international assistance, and SADC PF addresses this fact in draft benchmark 4.4.3(b), which states that "the type of assistance, the budget, and the use of these funds shall be determined in a transparent and accountable manner within parliament's strategic plan." Similarly to the Pacific version of the CPA benchmarks, SADC PF draft benchmark 4.3.1(b) provides specifically for parliamentary oversight of "all independent governmental bodies and constitutional bodies such as the human rights commission, ombudsman, director of public prosecutions, and public protector, among others." Draft benchmarks 4.3.2(b) and (c) also provide for parliamentary approval of "presidential appointments for offices that are of a nonpartisan nature. These include human rights commissioners, the ombudsman, electoral commissioners, auditor general, director of public prosecutions, and public protector, among others." Furthermore, the "President shall not remove these officials without notification and approval of parliament."

Among some of the other key changes or additions are the following:

- *1.1(e)* Dress codes in parliaments shall be culturally inclusive and shall not unduly limit public access.
- *4.1.1(c)* Parliaments shall approve all grants, loans, and guarantees, both domestic and international.
- *4.1.1(d)* Parliaments shall approve all treaties, protocols, and conventions.
- *4.2.1(d)* In addition, parliaments shall have a parliamentary budget office with a staff qualified to assist in budget analysis and monitoring of budget implementation on at least a quarterly basis.
- *4.3.2(a)* Parliaments shall enact a law to guarantee the right of parliament to create commissions of inquiry. Such commissions shall have the power to compel government officials to appear and give evidence under oath.
- *6.1.1(d)* Nominated or appointed members of parliament shall comprise not more than 5 percent of the overall size of the parliament.
- *6.1.1(g)* There shall be a minimum education requirement to determine eligibility to stand for parliament established by law in accordance with national standards, provided that where a candidate has relevant experience, the education requirement may be waived.
- *6.1.1(h)* Parliaments shall be take appropriate measures to assist members of parliament to increase their educational qualifications.
- *6.1.2(c)* Nomination fees shall be reasonable and affordable so as not to unduly exclude potential candidates.
- *6.1.2(f)* Parliaments shall enact laws that require political parties to take measures of affirmative action for gender in order to meet the provisions of the SADC Protocol on Gender and Development.

Table 1E.1 provides a comparative glimpse of the CPA, NDI, APF, and SADC PF standards or benchmarks under the general heading of elections, with new standards in *italics*.

Table 1E.1 Comparative Table of Standards or Benchmarks in the Category "Elections"

Organization	CPA	NDI	APF	SADC PF
Category	General	Election and Status of Legislators	Elections and Status of Legislators	Elections and Status of Members of Parliament
Subcategory	Elections	The Election of Legislators	Elections	Parliamentary Elections
Standard or benchmark	1.1.1 Members of the popularly elected or only house shall be elected by direct universal and equal suffrage in a free and secret ballot.	1.1.1 Members of the popularly elected or only house shall be directly elected through universal and equal suffrage in a free and secret ballot.	1.1.1 The national constitution shall include basic rules regarding elections and the status of legislators.	(a) Parliament shall enact all necessary laws to establish an independent electoral management body and to ensure free, fair, and credible elections.
Standard or benchmark	1.1.2 Legislative elections shall meet international standards for genuine and transparent elections.	1.1.2 Legislative elections shall meet international standards for genuine and transparent elections.	1.1.2 Legislators shall be elected by universal suffrage through an electoral process that is free, reliable, transparent, and in accordance with international and national standards. However, in a bicameral legislature, the second chamber may be governed by special regulations stipulated in the constitution or the laws of each country.	(b) Members of parliament shall be directly elected through universal and equal suffrage in a free and secret ballot in accordance with regional norms and standards for elections.
Standard or benchmark	1.1.3 Term lengths for members of the popular house shall reflect the need for accountability through regular and periodic legislative elections.	1.1.3 Term lengths for members of the popular house shall reflect the need for accountability through regular and periodic legislative elections.	1.1.3 Elections shall be held at regular intervals. Term limits shall be established for legislatures, and at the end of the term, new elections shall be held.	(c) Elections shall be held regularly and periodically.
Standard or benchmark			1.1.4 Elections shall take place without any restriction or violation of the freedom, security of person, freedom of opinion and speech, freedom of assembly and demonstration, and freedom of association of all voters and electoral candidates.	(d) Nominated or appointed members of parliament shall compose not more than 5 percent of the overall size of the parliament.

table continues next page

Table 1E.1 Comparative Table of Standards or Benchmarks in the Category "Elections" *(continued)*

Organization	CPA	NDI	APF	SADC PF
Standard or benchmark			1.1.5 *The organization and management of elections, from the preparatory procedures and electoral campaign to vote tallying and the announcement of results, shall be the responsibility of bodies vested with the authority to closely monitor the electoral process, ensure the credibility of the elections and the full participation of citizens in them, and ensure equal treatment of candidates throughout the electoral procedures.*	(e) *The selection of members of parliaments for reserved seats allocated for special groups shall be based on nonpartisanship.*
Standard or benchmark				(f) *The main legislative function shall be exercised by the directly elected chamber. Where a second chamber exists, such house shall have a secondary role.*
Standard or benchmark				(g) *There shall be a minimum education requirement to determine eligibility to stand for parliament established by law in accordance with national standards, provided that where a candidate has relevant experience, the education requirement may be waived.*
Standard or benchmark				(h) *Parliament shall take appropriate measures to assist members of parliament in increasing their educational qualifications.*

Sources: APF 2009; CPA 2006; NDI 2007; SADC PF 2010.
Note: New standards are in italics.

Notes

1. The terms *parliament* and *legislature* are used interchangeably in this chapter.

2. See, for example, the international election standards developed by International IDEA (2002) and the United Nations' Declaration of Principles for International Election Observation and Code of Conduct for International Election Observers (United Nations 2005), which was developed through a multiyear process involving more than 20 intergovernmental and international nongovernmental organizations, including the Commonwealth Parliamentary Association, the Inter-Parliamentary Union, the National Democratic Institute for International Affairs, and the Southern African Development Community Parliamentary Forum, among others.

3. For more about the World Bank's Parliamentary Strengthening Program, see the World Bank's website at http://www.worldbank.org/wbi/governance/parliament.

4. For more about UNDP's Global Program for Parliamentary Strengthening, see the UNDP's website at http://www.undp.org/content/undp/en/home/ourwork /democraticgovernance/focus_areas/focus_parliamentary_dev.html.

5. For example, the Parliamentary Assembly of the Council of Europe (PACE) may be doing so at present. According to draft minutes of a meeting held in London on December 7, 2009, the PACE Committee on Rules of Procedure, Immunities, and Institutional Affairs is looking at ways to "assess the strengths and weaknesses of Parliaments and to elaborate on this basis a model for an exemplary Parliament."

6. The quotation is displayed on NDI's website. For more information about the organization, visit http://www.ndi.org.

7. Originally founded as the Empire Parliamentary Association in 1911, the association took its current name, the Commonwealth Parliamentary Association, in 1948. The quotations in this paragraph are from the CPA's website. For more information about the association, go to http://www.cpahq.org.

8. The CPA thus differs from organizations such as the IPU, which does not have provincial parliaments as members.

9. The Parliamentary Study Group included parliamentarians from Bermuda, Canada, Ghana, Pakistan, and Scotland.

10. Several recommendations were taken specifically from the CPA (2005).

11. This description is in accordance with the APF's website. Visit the website at http:// apf.francophonie.org.

12. Examples of OIF work include the Bamako Declaration of November 2000 on democratic practices, rights, and freedoms in the Francophone world (http://www2.ohchr .org/english/law/compilation_democracy/oif.htm) and the St. Boniface Declaration of May 2006 issued by the Ministerial Conference of Francophone Countries on conflict prevention and human security (http://www.francophonie.org/IMG/pdf/Declaration _Saint-Boniface.pdf).

13. This description is in accordance with the IPU's website. Visit the website at http:// www.ipu.org.

14. Other examples of interest include the Parliamentary Powers Index (developed by M. Steven Fish and Matthew Kroenig); the Congressional Capabilities Index (developed by the International Development Bank); the International Foundation for Electoral Systems' Annual State of the Parliament Report; legislative strengthening indicators developed by the UNDP in 2001 and by other donors such as the U.S. Agency for International Development; the African Legislatures Project indicators;

Democracy Reporting International's Standards for Democratic Governance; Australian National University's Democratic Audit; the Arab Center for the Development of the Rule of Law and Integrity's Parliament/Participation Integrity Principles; and tools developed by civil society organizations in India, Pakistan, and Uganda.

15. More information on International IDEA's State of Democracy Assessment Methodology can be found at http://www.idea.int/sod/framework/.

16. The 13 pillars are the legislature, the executive, the judiciary, the public sector, law enforcement, the electoral management body, the ombudsman, the audit institution, anticorruption agencies, political parties, media, civil society, and business.

17. For more information, see Reid (2008).

18. The full text of the statement is available at https://www.ndi.org/files/Benchmarks _Conference_Participant_Statement_March2010.pdf.

19. The full text of the benchmarks is available at http://www.osisa.org/sites/default/files /sadc_parliamentary_forum_benchmarks_for_democratic_legislatures.pdf.

20. NDI and the APF recognize the power to lift immunity as exclusive to the parliament itself.

21. The CPA's parliamentary study group referred to its Zanzibar study group on the financing administration of parliament's recommendation that "The Corporate Body should ensure that the parliamentary service is properly remunerated and that retention strategies are in place" (CPA 2005).

22. Annex 1 of the IPU toolkit also notes the availability of nonpartisan professional staff members separate from the main civil service as a possible procedural and institutional means for effective organization of business (IPU 2008, 26).

23. CPA benchmark 2.4.2, NDI standard 2.3.2, and APF criterion 2.1.5.6 also cover elected legislators' right to initiate legislation and offer amendments to proposed legislation.

24. Along these lines, the different standards state that committee hearings shall be in public except in clearly defined circumstances that are provided for in the rules of procedure. Exceptions may include committee administration procedures, meetings where sensitive material related to national security is being reviewed, and witness protection situations (CPA 2006, 31).

25. For further discussion on committees, see, for example, NDI (1996) and Shaw (1998).

26. Very small parliaments may choose to work through a committee of the whole.

27. The CPA and the APF also developed standards relating to the use of multiple languages. See CPA benchmark 9.2.1 and APF criterion 4.1.3.1.

28. The survey typically takes each participant 15 to 20 minutes to complete, although in some cases NDI has used a guided interview methodology to administer the survey.

29. Participating organizations in this first meeting included the Canadian International Development Agency; Canadian Parliamentary Centre; Centre for Democratic Institutions; Constitution Unit of University College, London; CPA; Inter-American Development Bank; International Foundation for Electoral Systems; IPU; National Council of State Legislatures; NDI; Parlatino; UNDP; U.S. Agency for International Development; U.S. State Department, and World Bank.

30. The IPU, CPA, NDI, and others participate actively in this steering committee (or working group, as it is sometimes called).

31. The conference was organized by the World Bank and UNDP in partnership with the French Ministry of Foreign and European Affairs, European Parliament Office for the Promotion of Parliamentary Democracy, APF, CPA, IPU, and NDI and convened in Paris on March 2–4, 2010.

32. For example, in Samoa, an independent Salaries Tribunal decides on the remuneration of parliamentarians (and government officials). In Fiji and Kiribati, an independent body recommends salaries, but the final amount awarded has to be approved by the parliament.

33. The Tuvalu parliament's oversight function was considered ineffective, because the cabinet included more than half of the legislature's total membership.

34. Benchmark 4.3.1 states, "Legislators shall have the right to form interest caucuses around issues of common concern." An example of an interest caucus might be the Congressional Black Caucus in the U.S. Congress or any number of all-party groups in the U.K. Parliament.

35. The SADC PF draft benchmarks are available in English and Portuguese.

References

APF (Assemblée Parlementaire de la Francophonie). 2009. "La réalité démocratique des Parlements: Quels critères d'évaluation?" Text adopted by the 35th session of the APF, Paris, July. http://apf.francophonie.org/IMG/pdf/La_realite_democratique_des _Parlements_-_Quels_criteres_devaluation_-_Geneve.pdf.

Beetham, David. 2006. *Parliament and Democracy in the Twenty-First Century: A Guide to Good Practice*. Geneva: Inter-Parliamentary Union. http://www.ipu.org/PDF /publications/democracy_en.pdf.

CPA (Commonwealth Parliamentary Association). 2005. "The Administration and Financing of Parliament." Report of a CPA study group hosted by the Legislature of Zanzibar, Tanzania, March 25–29, CPA and World Bank, London.

———. 2006. "Recommended Benchmarks for Democratic Legislatures." CPA, London. http://wbi.worldbank.org/wbi/Data/wbi/wbicms/files/drupal-acquia/wbi /Recommended%20Benchmarks%20for%20Democratic%20Legislatures.pdf.

———. 2009. "CPA Benchmarks for Democratic Legislatures: Self-Assessment Guidance Note," CPA, London. http://www.ipu.org/splz-e/asgp09/dscr-CPA.pdf.

———. 2010. "Recommended Benchmarks for Asia, India, and South-East Asia Regions' Democratic Legislatures." CPA, London. http://www.agora-parl.org /sites/default/files/CPA%20Recommended%20Benchmarks%20for%20 Asia%2C%20India%20and%20South%20East%20Asia%20Regions%20 Democratic%20Legislatures.pdf.

CPA (Commonwealth Parliamentary Association), Commonwealth Legal Education Association, Commonwealth Magistrates' and Judges' Association, and Commonwealth Lawyers' Association. 2004. "Commonwealth (Latimer House) Principles on the Three Branches of Government." Commonwealth Secretariat, London.

Fish, M. Steven. 2006. "Stronger Legislatures, Stronger Democracies." *Journal of Democracy* 17 (1): 5–20.

International IDEA (International Institute for Democracy and Electoral Assistance). 2002. *International Electoral Standards: Guidelines for Reviewing the Legal Framework of Elections*. Stockholm: International IDEA.

IPU (Inter-Parliamentary Union). 2008. "Evaluating Parliament: A Self-Assessment Toolkit for Parliaments." IPU, Geneva. http://www.ipu.org/pdf/publications/self-e.pdf.

———. 2009a. "Carrying out A Self-Assessment: Preparation Note for Parliaments." IPU, Geneva.

———. 2009b. "Evaluating Parliament: A Self-Assessment Toolkit for Parliaments." Note prepared for the Inter-Parliamentary Union–Association of Secretaries General of Parliament Conference on Evaluating Parliament: Objectives, Methods Results, and Impact, Geneva, October 22. http://www.ipu.org/splz-e/asgp09/dscr-IPU.pdf.

LSE (London School of Economics and Political Science). 2009. *Parliamentary Assessment: An Analysis of Existing Frameworks and Application to Selected Countries*. London: LSE for the World Bank Institute.

NDI (National Democratic Institute of International Affairs). 1996. "Committees in Legislatures: A Division of Labor." Legislative Research Paper 2, NDI, Washington DC.

———. 2007. "Toward the Development of International Standards for Democratic Legislatures: A Discussion Document for Review by Interested Legislatures, Donors and International Organizations." NDI, Washington DC. http://www.ndi.org/files/2113_gov_standards_010107.pdf.

———. 2009. "National Democratic Institute for International Affairs (NDI) Survey on the Gaps between Parliamentary Power and Practice: Experiences in Colombia, Guatemala, Peru and Serbia." Note prepared for the Inter-Parliamentary Union–Association of Secretaries General of Parliament Conference on Evaluating Parliament: Objectives, Methods, Results, and Impact, Geneva, October 22.

OECD (Organisation for Economic Co-operation and Development). 2005. *Security System Reform and Governance*. Paris: OECD. http://www.oecd.org/development/incaf/31785288.pdf.

PACE (Parliamentary Assembly of the Council of Europe). 2009. "Self-Evaluation by Europe's National Parliaments: Procedural Guidelines." Strasbourg, France, PACE.

PILDAT (Pakistan Institute of Legislative Development and Transparency). 2009. "State of Democracy in Pakistan: Evaluation of Parliament, 2008–2009. PILDAT, Islamabad.

Reid, Gary J. 2008. "Actionable Governance Indicators: Concepts and Measurement." World Bank, Washington, DC. http://siteresources.worldbank.org/EXTPUBLICSECTORANDGOVERNANCE/Resources/286304-1235411288968/AGINote.pdf?resourceurlname=AGINote.pdf.

SADC PF (Southern African Development Community Parliamentary Forum). 2010. "Benchmarks for Democratic Legislatures in Southern Africa." SADC PF, Windhoek. http://www.agora-parl.org/node/2777.

Shaw, Malcolm. 1998. "Parliamentary Committees: a Global Perspective." *Journal of Legislative Studies* 4 (1): 225–51.

United Nations. 2005. "Declaration of Principles for International Election Observation and Code of Conduct for International Election Observers." United Nations, New York. https://www.ndi.org/files/1923_declaration_102705_0.pdf.

von Trapp, Lisa. 2007. "Donor Consultation on Parliamentary Development and Financial Accountability." U.K. Department for International Development, United Nations Development Programme, and World Bank Institute, Brussels.

———. 2010. "Benchmarks and Self-Assessment Frameworks for Democratic Parliaments." United Nations Development Programme, Brussels.

CHAPTER 2

The IPU's Self-Assessment Toolkit

David Beetham

Introduction

The Inter-Parliamentary Union (IPU) designed the Self-Assessment Toolkit for Parliaments to help parliamentarians conduct a systematic analysis of the performance of individual parliamentarians and parliaments as a whole (IPU 2008). The toolkit offers a framework to identify their main strengths and weaknesses against widely accepted criteria for democratic parliaments. From this assessment, parliamentarians can formulate priorities for improvement and assess the effectiveness of reforms already in progress.

Although the toolkit is organized as a series of assessment questions, it is not intended as a standard questionnaire to report to an external agency. Rather, the questions aim to facilitate discussion among parliamentarians, to explore differences of perception and judgment, and to foster agreement on priorities for change and improvement. Because the impetus for change has to come from within a parliament, any process of assessment is best owned and conducted by parliamentarians themselves. The toolkit provides members with the opportunity to stand back from their day-to-day work and reflect on their parliament's work in a systematic way.

The IPU developed the toolkit as part of a major program of work to examine what makes a parliament democratic, both in the way it functions and interacts with its electorate and in its effectiveness in performing its roles within a democratic system of government. Under this program, all IPU members were invited to contribute examples from their own reform experiences, which were compiled into a handbook of good practice titled *Parliament and Democracy in the Twenty-First Century* (Beetham 2006).[1] This exercise also informed the toolkit's key features, including its structure, its emphasis on parliament as the key site of a country's democracy, and its participant-led approach to assessing performance.

This chapter aims to review uses of the IPU's toolkit and is organized as follows: The first section reviews possible contexts for using the toolkit. The second section summarizes issues covered in the self-assessment process. The third section discusses how to use the toolkit. The fourth section highlights cases of countries that have used the toolkit. The final section concludes with next steps.

Possible Contexts for Using the Toolkit

The toolkit has the flexibility to be used in a variety of contexts:

- Responding to public concerns about the standing of parliament or specific aspects of its work
- Drawing up a strategic plan for parliament, including budget priorities
- Assisting an ongoing reform or modernization program, including the assessment of past reforms
- Working with a donor organization on a needs assessment to identify priorities for capacity building
- Conducting an assessment of parliament from a gender perspective

While the toolkit's precise format is context specific, several elements are common to all contexts. In particular, parliamentarians themselves should initiate the use of the toolkit. Moreover, key parliamentarians (such as members of an existing modernization committee) should be involved in the self-assessment. Furthermore, the assessment group should reflect the broadest range of perspectives within a parliament. Finally, the assessment's outcome should be a report with clearly identified priorities for action.

Issues Covered by the Toolkit

The toolkit is organized into six sections, which correspond to the key features of a democratic parliament and its roles in a democratic system of government:

- The representativeness of parliament
- Parliamentary oversight of the executive
- Parliament's legislative capacity
- The transparency and accessibility of parliament
- The accountability of parliament
- Parliament's involvement in international policy

Each section is designed to be self-standing, albeit within a coherent and interrelated whole, and comprises a series of questions, framed in a comparative way: "How far?" "How adequate?" "How effective?" The assumption behind this way of posing questions is that reaching a good standard of performance in any feature of a parliament's work is a matter of degree, not an all-or-nothing affair. In this way, the toolkit differs from a checklist of indicators to which a simple "yes" or "no" response is expected. Members are invited to provide a provisional score of their parliament's performance under each question on a five-point scale, from "very high or very good" to "very low or very poor."

In each of the six sections, comparative questions are followed by three qualitative questions: "What has been the biggest recent improvement in the above?" "What is the most serious ongoing deficiency?" "What measures would be needed to remedy this deficiency?" These questions emphasize that what is important is the substantive issue that lies behind the numerical scores. Space is also provided for members to add additional questions if they feel that those provided do not cover all the relevant issues facing their particular parliament.

As an example, one section of the toolkit is given in box 2.1. The full list of questions under all six sections can be found in annex 2A.

Using the Toolkit

Readers of this chapter are invited to score their own parliament by answering the questions in box 2.1. This is the best way to get a feel for the toolkit, to understand how it works, and to anticipate any difficulties that may arise. Although readers will undertake this exercise on their own, the intention of the self-assessment process is that scoring the questions should be done as part of a group. It is thus suggested that participants work out their own scores initially but then open up a collective discussion of the issues that lie behind the scores. In some cases, significant differences of opinion may arise among group members about these preliminary assessments. The discussion should aim to identify the reasons for these differences, to reconcile them where possible, and to record them as part of the assessment reporting process.

Box 2.1 Toolkit Example

Following is a portion of the questionnaire used by parliaments to assess their performance. A chart that allows members to mark their responses, from "very high or very good" to "very low or very poor," follows the questionnaire. After the chart are three qualitative questions.

5. The accountability of parliament

  5.1 How systematic are arrangements for members to report to their constituents about their performance in office?

  5.2 How effective is the electoral system in ensuring the accountability of parliament, individually and collectively, to the electorate?

  5.3 How effective is the system for ensuring the observance of agreed codes of conduct by members?

  5.4 How transparent and robust are the procedures for preventing conflicts of financial and other interests in the conduct of parliamentary business?

  5.5 How adequate is the oversight of party and candidate funding to ensure that members preserve independence in the performance of their duties?

  5.6 How publicly acceptable is the system whereby members' salaries are determined?

---

box continues next page

Box 2.1 Toolkit Example *(continued)*

5.7 How systematic are the monitoring and review of levels of public confidence in parliament?

5.8 Additional questions:

	Very high or very good	*High or good*	*Medium*	*Low or poor*	*Very low or very poor*
	5	**4**	**3**	**2**	**1**
5.1					
5.2					
5.3					
5.4					
5.5					
5.6					
5.7					
5.8					

What has been the biggest recent improvement in the above? _____
_____
_____

What is the most serious ongoing deficiency? _____
_____
_____

What measures would be needed to remedy this deficiency? _____
_____
_____

Before a self-assessment is carried out, however, parliaments should consider several questions:

- What is the purpose of the self-assessment? Does everyone involved share the same understanding?
- What is the expected outcome of the exercise?
- Who will participate in the self-assessment? Does the group represent a broad range of perspectives in parliament?
- Will the group interact with people outside parliament? If so, how will these interactions take place?
- What outcome documents will be produced? How will they be used? To whom and how will they be disseminated?

- Have sufficient resources been allocated to the self-assessment?
- Has a realistic timeframe been established for the exercise?

The rest of this section reviews frequently asked questions relating to the toolkit, discusses other partners that should be involved in the assessment process, and highlights the importance of an outcome document.

Frequently Asked Questions

The following questions raised by toolkit users in past conferences, seminars, pilot projects, or full assessments offer useful background about the toolkit's purpose and use:

- *How do we know what counts as a "good" or "very good" standard in relation to a particular question?* Comparative knowledge of good practice from other parliaments is useful, though established members will already have a fair idea of where their parliament's practice may be deficient. In assessing an existing reform process, a comparison with the parliament's own past may be relevant. Moreover, an internal comparison between the different components in a given section helps develop sensitivity to what is done better or not as well.

- *What if, in answering a particular question, we find both good and poor features, so that a "medium" score can be misleading?* The individual scores form only the starting point of the exercise, and what lies behind a given score should be brought out in discussion. When the toolkit was developed, questions were not broken down into all possible subcomponents because of concern that a huge list would be off-putting.

- *Is it realistic to expect that a parliament could score well on every component?* In democracy analysis, there is the well-known phenomenon of trade-offs. Not all good features can be maximized simultaneously. For example, a proportional electoral system may produce a more politically representative parliament but reduce the accessibility or accountability of individual members to their constituents.

- *Are the questions equally suitable for every type of parliament or parliamentary system?* The toolkit should be relevant for all parliaments. However, not every question may apply, and the opportunity to insert additional questions allows the distinctive features or concerns of a particular parliament to be included.

- *How long should it take to complete a single section?* Each individual can complete a section relatively quickly on his or her own. However, the most important thing is the collective discussion that follows, and no uniform time can be prescribed for this discussion. It is recommended that no more than one section be completed in a given sitting.

- *What if time is limited?* Parliamentarians typically have limited time available. Hence, it may be necessary to divide the toolkit sections among different sub-groups of a self-assessment committee according to interest or to concentrate on some sections only, depending on the overall purpose of the self-assessment exercise. The ownership and quality of the process may be as important as output or quantity. However, it is essential that all participants in the process have a broad overall view of the toolkit sections and their interconnectedness, even if they are asked to concentrate on particular areas within it.

- *Should the scores for each question be aggregated to provide an overall score, either for each section or even for the parliament as a whole?* The public loves arithmetical scores, and parliaments such as the Pakistani parliament have done this. The important thing is the identification of specific strengths and weaknesses through a discursive process, and the scores are merely the starting point for this discussion. The danger is that a process of aggregation can mask the differentiated character of a parliament's performance—good in some features, not as good in others. Score aggregation could also invite a comparison between parliaments, or even the construction of a league table of performance (ranking chart), which is not the purpose of the toolkit.

- *What kind of support does the IPU provide throughout the self-assessment process?* The IPU has published an explanatory booklet titled "Evaluating Parliament: A Self-Assessment Toolkit for Parliaments" (IPU 2008), which parliamentarians are recommended to read before undertaking a self-assessment. The booklet, which is available in English, French, Spanish, and Arabic, describes the basic features and purpose of the toolkit and provides guidance on how to use it. The IPU also provides an advisory service to parliaments wishing to carry out a self-assessment exercise, as well as trained facilitators to assist the process. The toolkit has been incorporated into the IPU's technical cooperation program and is used in carrying out a needs assessment so as to place parliament's needs at the center of parliamentary strengthening projects and to ensure parliamentary ownership.

Who Else Might Be Involved in an Assessment?

Because the improvement of parliament is a matter of wider public concern, the fact that parliament is engaged in a self-assessment could potentially arouse much public interest. Those conducting an assessment may also find it useful to invite nonparliamentarians to contribute to the process at an appropriate stage, depending on the particular purpose and focus of the assessment. For example, academic specialists in legislative affairs could bring a comparative perspective to bear from their knowledge of other parliaments, opinion-polling experts could provide a more detailed understanding of public attitudes toward parliament, journalists could share the media's perspective on parliamentary effectiveness, and members from women's rights nongovernmental organizations could help strengthen a gender perspective. Involving outsiders is a matter of discretion, according to the context and time available.

An Outcome Document

The outcome of the self-assessment process should be a report of the discussions and conclusions reached. For each section of the toolkit, the report might include the following:

- The main strengths of the parliament
- The areas requiring significant improvement
- Possible institutional means for realizing these improvements
- Potential obstacles and how they might be overcome
- Any significant differences among members in their responses

The outcome document should preferably be presented to a full plenary of parliament, as well as to the relevant committees responsible for taking the development process forward.

Consideration should also be given to whether the findings should be published more widely at the end of the process and how a media strategy should be developed. The fact that parliament is undertaking a self-assessment could have a positive effect on the public's perception of parliament. The opportunity could be used to explain the range of parliament's work and its contribution to democracy. Moreover, those working to improve parliament can benefit from the support of key individuals and groups in civil society.

Use of the Toolkit to Date

Parliamentarians have tested the toolkit in a number of international seminars and conferences before its use by parliaments themselves. To date, it has been used in a variety of countries and contexts, as table 2.1 shows. Some of these country experiences are the subject of later chapters.

Table 2.1 Examples of Countries That Have Used the Self-Assessment Toolkit

	Who initiated	Purpose	Process
Andorra	Initiated by parliament	To gather recommendations to improve parliamentary performance	Seminar with parliamentarians
Cambodia	Initiated by parliament	To review the Senate after 10 years of existence	Seminar with parliamentarians, academics, media, and civil society
Ireland	Initiated by parliament, through the secretary general's office, and piloted by the Library and Research Service	To identify potential reform areas and to use the results for a long-term planning and vision exercise for the Oireachtas	Focus groups of members of the Oireachtas Commission, committees of both chambers, and the Informal Feedback Forum

table continues next page

Table 2.1 Examples of Countries That Have Used the Self-Assessment Toolkit *(continued)*

	Who initiated	*Purpose*	*Process*
Pakistan	Initiated by parliament; facilitated by the Pakistan Institute of Legislative Development and Transparency	To gather recommendations to improve parliamentary performance	Creation of a questionnaire with objective and subjective indicators; validation conference to review results
Rwanda	Initiated by parliament; facilitated by the Inter-Parliamentary Union	To review parliament's strategic plan	Creation of an ad hoc committee in each chamber to carry out the self-assessment
Sierra Leone	Initiated by the Inter-Parliamentary Union as part of a technical cooperation project	To create a strategic plan	Seminars with parliamentarians
Thailand	Initiated by King Prajadhipok's Institute	To gather recommendations to improve parliamentary performance	Creation of an index to evaluate the performance of the Thai National Assembly

Conclusion and Next Steps

In summary, the IPU designed the self-assessment toolkit to assist parliamentarians in conducting a systematic analysis of the performance of individual parliamentarians and parliaments as a whole. Through the toolkit's process of identifying strengths and weaknesses, parliamentarians can formulate priorities for improvement and assess the effectiveness of reforms already in progress.

Besides continuing to actively promote the toolkit, the IPU is engaged in two further developments of the self-assessment principle:

- *Creation of a self-assessment tool to examine gender sensitivity in parliaments.* In 2011, the IPU published the results of a research project on gender-sensitive parliaments. The project has gathered primary information on the ways in which parliaments can best become gender-sensitive institutions and effectively mainstream gender in their work. Based on the experience with the self-assessment toolkit, the next step will be to develop guidelines for parliaments to assess their own gender sensitivity.

- *Creation of a voluntary review mechanism of parliamentary performance.* The mechanism would offer parliaments an opportunity to exercise collective responsibility and assist one another in assessing and improving their respective performance. Like similar mechanisms that have been established at the United Nations and regional organizations, the IPU exercise would be based on agreed values, codes, and criteria. The review process would be consultative, participatory, and transparent, as well as grounded in dialogue and interaction among key stakeholders. Participation in the process would be entirely voluntary, and the process in each case would be nationally owned.

Annex 2A: Full List of Self-Assessment Questions

1. The representativeness of parliament

    1.1 How adequately does the composition of parliament represent the diversity of political opinion in the country (for example, as reflected in votes for the respective political parties)?

    1.2 How representative of women is the composition of parliament?

    1.3 How representative of marginalized groups and regions is the composition of parliament?

    1.4 How easy is it for a person of average means to be elected to parliament?

    1.5 How adequate are internal party arrangements for improving imbalances in parliamentary representation?

    1.6 How adequate are arrangements for ensuring that opposition and minority parties or groups and their members can effectively contribute to the work of parliament?

    1.7 How conducive is the infrastructure of parliament and its unwritten mores to the participation of women and men?

    1.8 How secure is the right of all members to freely express their opinions, and how well are members protected from executive or legal interference?

    1.9 How effective is parliament as a forum for debate on questions of public concern?

2. Parliamentary oversight of the executive

    2.1 How rigorous and systematic are the procedures whereby members can question the executive and secure adequate information from it?

    2.2 How effective are specialist committees in carrying out their oversight function?

    2.3 How well is parliament able to influence and scrutinize the national budget through all its stages?

    2.4 How effectively can parliament scrutinize appointments to executive posts and hold their occupants to account?

    2.5 How far is parliament able to hold nonelected public bodies to account?

    2.6 How far is parliament autonomous in practice from the executive (for example, through control over its own budget, agenda, timetable, personnel, and so forth)?

    2.7 How adequate are the numbers and expertise of the professional staff to support members, individually and collectively, in the effective performance of their duties?

    2.8 How adequate are the research, information, and other facilities available to members and their groups?

3. Parliament's legislative capacity

    3.1 How satisfactory are the procedures for subjecting draft legislation to full and open debate in parliament?

3.2 How effective are committee procedures for scrutinizing and amending draft legislation?

3.3 How systematic and transparent are the procedures for consultation with relevant groups and interests in the course of legislation?

3.4 How adequate are the opportunities for individual members to introduce draft legislation?

3.5 How effective is parliament in ensuring that legislation enacted is clear, concise, and intelligible?

3.6 How careful is parliament in ensuring that legislation enacted is consistent with the constitution and the human rights of the population?

3.7 How careful is parliament in ensuring a gender-equality perspective in its work?

4. The transparency and accessibility of parliament

4.1 How open and accessible to the media and the public are the proceedings of parliament and its committees?

4.2 How free from restrictions are journalists in reporting on parliament and the activities of its members?

4.3 How effective is parliament in informing the public about its work through a variety of channels?

4.4 How extensive and successful are attempts to interest young people in the work of parliament?

4.5 How adequate are the opportunities for electors to express their views and concerns directly to their representatives, regardless of party affiliation?

4.6 How user-friendly is the procedure for individuals and groups to make submissions to a parliamentary committee or commission of inquiry?

4.7 How much opportunity do citizens have for direct involvement in legislation (for example, through citizens' initiatives and referenda)?

5. The accountability of parliament

5.1 How systematic are arrangements for members to report to their constituents about their performance in office?

5.2 How effective is the electoral system in ensuring the accountability of parliament, individually and collectively, to the electorate?

5.3 How effective is the system for ensuring the observance of agreed codes of conduct by members?

5.4 How transparent and robust are the procedures for preventing conflicts of financial and other interests in the conduct of parliamentary business?

5.5 How adequate is the oversight of party and candidate funding to ensure that members preserve independence in the performance of their duties?

5.6 How publicly acceptable is the system whereby members' salaries are determined?

5.7 How systematic are the monitoring and review of levels of public confidence in parliament?

6. Parliament's involvement in international policy

   6.1 How effectively can parliament scrutinize and contribute to the government's foreign policy?

   6.2 How adequate and timely is the information available to parliament about the government's negotiating positions in regional and international bodies?

   6.3 How much can parliament influence the binding legal or financial commitments made by the government in international forums, such as the United Nations?

   6.4 How effective is parliament in ensuring that international commitments are implemented at the national level?

   6.5 How effectively can parliament scrutinize and contribute to national reports to international monitoring mechanisms and ensure follow-up on their recommendations?

   6.6 How effective is parliamentary monitoring of the government's development policy, whether as "donor" or "recipient" of international development aid?

   6.7 How rigorous is parliamentary oversight of the deployment of the country's armed forces abroad?

   6.8 How active is parliament in fostering political dialogue for conflict resolution, both at home and abroad?

   6.9 How effective is parliament in interparliamentary cooperation at the regional and global levels?

   6.10 How much can parliament scrutinize the policies and performance of international organizations such as the United Nations, the World Bank, and the International Monetary Fund, to which its government contributes financial, human, and material resources?

Note

1. The handbook serves as a useful reference point for identifying standards and examples of good practice in the course of carrying out a self-assessment.

References

Beetham, David. 2006. *Parliament and Democracy in the Twenty-First Century: A Guide to Good Practice.* Geneva: Inter-Parliamentary Union. http://www.ipu.org/PDF /publications/democracy_en.pdf.

IPU (Inter-Parliamentary Union). 2008. "Evaluating Parliament: A Self-Assessment Toolkit for Parliaments." IPU, Geneva. http://www.ipu.org/pdf/publications/self-e.pdf.

CHAPTER 3

Benchmarks for Commonwealth Parliaments and Legislatures

Akbar Khan

Background

In the latter part of the 20th century, two Commonwealth small-island parliaments encountered problems with removing speakers of house who chose to ignore the political reality that they no longer held the confidence of their respective houses and the convention that they should leave office immediately. The mechanisms and precedents that each house had to remove its speaker were clumsy and difficult to implement, and applying them would have taken a protracted time that their parliaments and their countries could not afford. Effective and appropriate rules for removing a recalcitrant speaker had not been developed because members—particularly speakers—had understood the convention that a presiding officer who had lost the confidence of the house should step down, preferably before the house had to express its displeasure publicly. Both situations were ultimately resolved when the people at the center of the disputes finally exhausted any legal avenues they thought they had and gave up their fights. Over the centuries, these and other occurrences, both common and unusual, were consigned to precedent in the standing orders or procedural guides in the relevant parliament. News of developments was shared among the 185 Commonwealth parliaments and legislatures and became part of a general understanding of how parliament should work. These developments were never consolidated into one common standard of good Commonwealth parliamentary practice.

Throughout the century of its existence, the Commonwealth Parliamentary Association (CPA) and its predecessor, the Empire Parliamentary Association, have promoted this type of understanding as they enabled parliamentarians and parliamentary officials to share information and experiences. An understanding of and a respect for the broad tenets of parliamentary democracy and the diversity of practices and procedures among sovereign countries—two of the

Mr. Akbar Khan is the Secretary-General of the Commonwealth Parliamentary Association (www.cpahq .org) from January 2016.

Commonwealth's basic principles—governed the sharing of experiences in the parliamentary community. Thanks to a common background in parliamentary governance, the rule of law, and respect for political rights, there never seemed to be a need to define precisely the basic elements of an ideal parliament or even a properly functioning one. Nor did the CPA or any individual parliament attempt to dictate how countries should be governed or how their parliaments and legislatures ought to be run. Members and clerks simply understood, without the need for codification. This approach served the evolution of parliamentary governance well as the association encouraged individual parliaments and members to adapt practices elsewhere to their own situations, thus creating new practices.

Development of the Benchmarks

This situation began to change toward the turn of the millennium as intergovernmental organizations, aid agencies, and donor governments realized that improving not just governments but also parliaments was essential when working to strengthen countries. Therefore, bodies that had years of experience with assessing the potential of administrative reforms to bureaucracies or the value of building hospitals or training teachers began to look for standards that would enable them to assess parliamentary assistance requests and propose parliamentary strengthening programs of their own.

In 2001, the United Nations Development Programme (UNDP) published its own parliamentary indicators. This development was followed in January 2004 with a discussion by the Australia and New Zealand Association of Clerks-at-the-Table of the use of benchmarks for parliamentary administration. In September 2004, parliamentarians meeting at the CPA's Commonwealth Parliamentary Conference in Canada agreed that because standards were being drafted and then used to assess parliaments, they should be drafted by parliamentarians and parliamentary officials with firsthand experience of how this highly specialized area of governance actually worked in theory and practice. Shopping lists of good governance ideas advocated by donor agencies, the executive arms of governments, academics, and interest groups, as well as masses of recommendations on good electoral practices, could not be allowed to take over parliamentary reform agendas without input from the people who were responsible for making the institutions work.

In December 2004, the CPA and the World Bank convened a meeting in Washington, DC, composed of about 20 donors, intergovernmental organizations, and interparliamentary organizations to discuss how best to develop benchmarks against which parliamentary assistance projects could be measured. It was eventually agreed that the CPA and the World Bank would hold a CPA study group so that a representative group of parliamentarians and clerks could draft a basic set of standards constituting good practice for a parliamentary democracy.

At that point, the CPA had used the study group process over many years, which enabled 26 expert groups of members and officials to recommend good practices in many individual aspects of parliamentary government ranging from the scrutiny of public finance, to the consideration of science policy, to the security of small states. More recently, some groups examined what constituted good relationships between parliament and the media; good practices in freedom of information, HIV/AIDS policies, and parliament's use of information and communications technologies; and good modern practices in parliamentary administration and finance. A group organized by the CPA and the Commonwealth's three legal professional associations drafted guidelines on the separation of powers among the executive, the judiciary, and parliament. These guidelines became known as the "Commonwealth (Latimer House) Principles on the Accountability of and Relationship between the Three Branches of Government" (CPA and others 2004) and were adopted in 2003 by Commonwealth heads of government to underpin Commonwealth governmental values, eventually becoming part of the 2013 Charter of the Commonwealth. The approach of allowing expert groups of members of parliament and clerks to recommend good practices enabled the association to respect the Commonwealth's commitment to diversity; that is, groups, not the full association, made the recommendations and stopped short of dictating what should be best practice to be followed by all. For some parliaments, production of recommendations in limited areas of parliamentary democratic practice was descriptive of current practice; for others, it was aspirational, serving as a suitable format for an examination of the parliamentary system as a whole.

The CPA and the World Bank therefore joined the Parliament of Bermuda in hosting a parliamentary standards study group in Hamilton in December 2006. The UNDP and the European Parliament also supported the meeting. Earlier in 2006, the National Democratic Institute for International Affairs (NDI) had published a set of standards based on the U.S. congressional system, so it too was invited to join the group. Its standards (a) assisted the study group in synthesizing some of the recommendations that had emerged from the CPA's earlier work and (b) helped codify previously unstated understandings into the "Recommended Benchmarks for Democratic Legislatures" (CPA 2006). In September 2007, the Commonwealth Parliamentary Conference in India discussed the benchmarks and their use in helping the donor community and in facilitating self-assessments by parliaments.

As the 2006 group noted, the benchmarks strengthen the cases that individual members, officials, and parliaments can make to convince others—especially governments—of the merits of a particular reform proposal. A benchmarks self-assessment can lead to discussion and debate—both inside and outside of parliaments—about their appropriateness and utility in different nations. They are useful tools around which to formulate a debate within a parliament as to the potential directions for parliamentary reform. The benchmarks help to leverage reforms because individual presiding officers, members, committees, parliamentary officers, or governments can make a stronger case for change by pointing to external standards, especially those set by representative groups of Commonwealth parliamentarians and officials who work in environments similar to their own.

Parliaments were used to being assessed constantly by outsiders—for example, the media, academics, intergovernmental agencies, civil society bodies, and individual citizens—usually with little understanding of how a parliament works. The CPA benchmarks finally enabled parliaments to assess themselves against a Commonwealth standard developed by members and parliamentary officials to help parliaments determine how they can function more effectively.

The benchmarks also brought parliaments into line with other institutions, industries, and professions that were adopting standards reflecting new thinking in management and the achievement of desired levels of productivity and quality as societies demanded ever higher levels of accountability, responsibility, and transparency from institutions, organizations, professions, and companies.

Content of the Benchmarks

The benchmarks divide parliamentary democracy into four main areas. The first part of the document details how members get into parliament and the general features of the institution. The next part recommends how the institution should be organized, operated, and administered, which is followed by a focus on the main functions of the legislature in making laws, overseeing the executive, and representing the views of the people. The document concludes with recommendations on public access, transparency, and integrity provisions to provide ethical governance. The 87 Commonwealth benchmarks are found in annex 3A.

The Commonwealth's diversity, which was a significant part of the reason the CPA avoided dictating "best practice" for nearly a century, also underpins the value of interparliamentary cooperation and consultation: if all parliaments followed the same practices and procedures, there would be fewer approaches to share for possible adoption or adaptation in other assemblies. A parliamentary system slavishly following one model would be extremely difficult to reform and would be unable to keep pace with advances in the Commonwealth's diverse societies. In fact, parliaments throughout the Commonwealth have always changed to meet new demands, to exploit new opportunities such as those presented by new technologies, and to meet ever-higher expectations on the part of citizens and parliamentarians and parliamentary officials themselves. The benchmarks contribute to this evolutionary process.

Development of Regional Benchmarks

The benchmarks were always intended to mark the beginning of the discussion rather than the end. The path to good parliamentary democracy clearly is far too complicated to be marked by a mere 87 signposts. Additionally, the study group represented most of the Commonwealth regions, but it could not capture all of the nuances and diversity in today's 53 Commonwealth countries. For example, how do you accommodate cultural, social, and religious differences that produce different social obligations—such as those of the Malaysian sultans, the Samoan matai, or the aboriginal minorities in Australia and Canada that must be respected or actually be placed in their systems of democratic representation and governance? If a parliament must reflect its society to be relevant to it, then so must standards of assessment if they are to be useful to that parliament.

Therefore, the CPA embarked on the development of regional benchmarks to codify differences in practices among broad cultural groups, to identify higher standards that some regions would be willing to set for themselves but which others might not yet be ready to accept, and to help to identify other important aspects of governance that would benefit from benchmarking. This effort was inspired by the Southern African Development Community Parliamentary Forum, which began a program in 2008 to modify the CPA benchmarks with a view to form a set of benchmarks for southern African parliaments that were acceptable to their combined Commonwealth, francophone, and lusophone traditions. The program culminated in November 2011 with the official adoption of the southern African benchmarks (SADC PF 2010).

The Pacific region was the first to adopt standards based on the CPA benchmarks. Benchmarks for small Pacific island legislatures were adopted in 2009 in a joint initiative of the CPA, the World Bank, UNDP, and the Pacific Legislatures for Population and Governance, with assistance provided by Australia's Centre for Democratic Institutions and by the parliaments of Australia, the Cook Islands, Kiribati, Nauru, New South Wales, Niue, Queensland, Tuvalu, and Vanuatu (CPA 2009).

Asian benchmarks for the CPA's Asian, Indian, and Southeast Asian regions were adopted in 2010 in a joint program with the UNDP and the World Bank, with support from the Bangladeshi parliament and other Commonwealth houses, including those of Malaysia, Pakistan, and Singapore (CPA 2010). The Caribbean followed with its benchmarks in 2011 as a purely regional exercise led by the CPA Regional Secretariat in Trinidad and Tobago as well as by parliamentary officials from Barbados, the Cayman Islands, Grenada, and Jamaica (CPA 2011). Also included was the entire Commonwealth Caribbean via regional meetings of speakers and clerks in the Cayman Islands and parliamentary delegations in Grenada. It was agreed by the Caribbean, Americas, and Atlantic (CAA) Region that each branch should report on its progress in seeking to attain the CAA Regional Benchmarks.

Going Forward

The benchmarks will not remain static, and the CPA continues to ensure their relevance by producing regional variations and by bringing into the process older parliaments in more developed countries and parliaments in other parts of the Commonwealth. The CPA also looks specifically at different areas not now covered in the current sets of benchmarks, such as the scrutiny of delegated legislation or the constituency development funds. It is also considering benchmarks for specific sectors of the parliamentary community, such as benchmarks for parliamentary administration or parliamentary information dissemination, or benchmarks for presiding officers or public accounts committees.

At a CPA study group meeting on "Improving Parliamentary Performance in a Tech-Enabled World" in London in May 2013—again held with NDI in attendance—new benchmarks were proposed that would set standards for the behavior of individual parliamentarians regarding personal conduct during parliamentary and political business. Many parliaments already have codes of conduct for financial behavior, but few go beyond that to other aspects of individual behavior in the political scene. Attendees of the meeting noted that even isolated instances of rowdy, abusive, or disruptive behavior by individual members of parliament in the chamber, in committees, or in other public settings can be blown out of proportion via social media and thereby discredit the institution of parliament. They agreed that the benchmarks should move beyond institutional performance to set standards for individual performance, again taking into account the differences in what is considered acceptable in different Commonwealth societies and at different times.

At the 60th Commonwealth Parliamentary Conference (CPC) in October 2014 in Yaoundé, Cameroon, a series of interviews were conducted on behalf of the CPA by senior academic staff from Monash University (Australia) and the University of Dar es Salaam (Tanzania) in relation to parliamentary codes of conduct applying to members of parliament across the Commonwealth. A workshop was also held at the Conference which considered 'Parliamentarians and public trust: do codes of conduct help?' One of the agreed recommendations from the workshop was to prepare a set of benchmarks to guide parliaments in the development of codes of conduct.

Following the 60th CPC, a workshop was convened in Victoria, Australia, in April 2015. Using a similar process to the original CPA Benchmarks for Democratic Legislature Parliamentarians, Senior Parliamentary Clerks, and other experts met to discuss the establishment of a Code of Conduct for individual parliamentarians.

These new Recommended Benchmarks for Codes of Conduct for Members of Parliament would complement the existing Recommended Benchmarks for Democratic Legislatures but would focus on individual Members, as opposed to the latter code, which looked at the Institution of Parliament.

The meeting saw the development of a set of Recommended Benchmarks for Codes of Conduct for Members of Parliament. These Benchmarks have been designed to be used by individual Houses of Parliament or other Legislatures to help them to revise and strengthen existing provisions affecting the conduct of their Members or to develop new codes of conduct.

The Benchmarks that have been developed are general in nature, so that they can be adapted to any parliament, ranging from small states and their assemblies to the largest, and from the least developed to the well resourced.

The CPA has encouraged all its Branches to use these Benchmarks as a set of provisions related to each other and together aimed to improve the integrity and performance of each Legislature; to take the underlying contribution to integrity of each recommended Benchmark and adapt it to a particular parliamentary system so as to guide the conduct of Members to benefit the performance of the Parliament.

The evolution of the benchmarks is not confined to what is and what is not recommended. The way they are used is also being debated and refined, and there are those who still question the value of trying to measure parliamentary performance. Some external assessment systems have contributed to these doubts by, for example, judging parliamentary performance on the basis of political output and assuming that passing more laws equates with better parliamentary performance. Because the initiation of new laws is primarily a government's role, is it relevant to judge the performance of a parliament on that basis? Even more debatable is the idea that more laws mean better governance.

Are measurements relevant at all in politics? In the United Kingdom, the Labour Party government went into the 2007 Scottish parliamentary elections with the ability to argue that it had implemented virtually all of the policies it had been elected to deliver. Nevertheless, it lost the election. What is the point of performance assessment in politics if one can tick all the boxes except the most important one—the one on the ballot paper? Do a parliament's clients care if it does all the right things? In British Columbia, the legislative assembly responded to public complaints in the 1980s and 1990s about the integrity of parliamentarians by introducing strict codes of conduct, rules on asset disclosures, and transparency requirements, yet public perceptions of the integrity of their political leaders continued to fall in public opinion polls.

Is it misleading to use the benchmarks in a box-ticking exercise? For example, a parliament that has many committees on paper but only a handful of committee clerks and no meeting rooms could correctly say that it has committees, even though in reality they do not contribute to the improved scrutiny of the executive. If governments are unable or refuse to provide parliaments with adequate resources, will this fact emerge in assessments?

Would it be better to rate each benchmark on a scale? Are some benchmarks more important than others, rating a higher maximum mark on a longer scale? Are different benchmarks more important to different countries or to the same country in different circumstances? The second benchmark calls on elections to meet internationally recognized standards, most of which include the stipulation that elections are to be run by an independent commission. The United Kingdom does not conform to that standard, but no one would question the fairness of its parliamentary elections, which are run by local authorities as—effectively—individual elections in each constituency. It is recommended that a parliament should control its own budget, but the Australian legislature does not, apparently with no long-term adverse effects on the quality of that parliament.

Although the CPA and its member parliaments and legislatures support the benchmarks as a self-assessment tool, there are differences of opinion on who should be part of a parliament's self-assessment. An assessment panel could include presiding officers, government and opposition members, clerks or secretaries, and other officials. However, the argument is made that only backbenchers should be involved and that the parliamentary staff would not be free to respond honestly (or at all) in the presence of members. The panel could have added credibility in the eyes of some—notably an increasingly cynical public—if it included some respected and knowledgeable external assessors, such as judges, senior civil servants, lawyers, academics, or former members of parliament or parliamentary officials.

In 2014, at the request of the CAA Region, the CPA secretariat convened an assessment workshop for the CAA Region to provide an opportunity for representatives from the region to self-assess their parliaments against the *CAA Regional Benchmarks*. The assessment workshop was held from 25–26 July 2014, in Bridgetown, Barbados, and included presiding officers, members, clerks and senior parliamentary staff.

The workshop also included resource personnel from NDI in addition to representatives from two Commonwealth Parliaments that had undergone a self-assessment of their institutions against the *CPA Benchmarks*. The parliaments were the Canadian Parliament and the Parliament of the Australia Capital Territory (ACT).

In advance of the assessment workshop, participants assessed their institutions against the *CPA Benchmarks for the Caribbean, Americas, and Atlantic* using a 5-point scale:

Fully meets the benchmark	5
Partially meets the benchmark	4
Currently developing processes to implement the benchmark	3
Reviewing potential application of benchmark	2
No current plan to meet benchmark	1

During the course of the two days, there was a relatively large degree of consistency within the CAA Region with respect to many of the *CAA Regional Benchmarks*. In general, participating Branches showed a clear and common understanding of the meaning and application of the Benchmarks. Many participating Branches felt that they met a large majority of the *CAA Regional Benchmarks* but shared similar challenges throughout the CAA Region. For example, it was noted that in many Parliaments in the Region, the number of the Members is sufficiently small that, once cabinet Members are excluded, there is sometimes an insufficient number of Members for a robust committee system. Accordingly, Members are required to serve on many committees and have challenges in addressing these competing committee mandates. This has the effect of reducing the effectiveness of committees in studying legislation, government operations, or specific topics.

The most pronounced differences among the participating Branches tended to involve differences between the British Overseas Territories and other jurisdictions. It was acknowledged that section 3 of the Regional Benchmarks, *"Functions of the Legislature" (governing the legislative, oversight, and representation functions of legislatures)*, posed certain challenges for many of these jurisdictions given their status.

Despite these differences of opinion—or perhaps because of them—the CPA plans to work with its partner organizations to reexamine the current Commonwealth benchmarks and the regional and sectoral versions to produce a new Commonwealth-wide standard. This effort will encourage parliaments and legislatures to aspire to even higher standards and reinforce the new Commonwealth Charter adopted by Commonwealth heads of government and formally signed by Her Majesty Queen Elizabeth II as head of the commonwealth in March 2013. Parliamentary benchmarks strengthen the core values and principles of the Commonwealth and therefore strengthen the foundation of this incredibly valuable global network of nations and people who share an understanding of governance that has more in common than any other group of nations.

Annex 3A: Recommended Benchmarks for Democratic Legislatures

The benchmarks that follow are the outcome of a 2006 study group hosted by the Parliament of Bermuda on behalf of the CPA and the World Bank with support from UNDP, the European Parliament, and NDI.

I. General

  1.1 Elections

    1.1.1 Members of the popularly elected or only house shall be elected by direct universal and equal suffrage in a free and secret ballot.

    1.1.2 Legislative elections shall meet international standards for genuine and transparent elections.

    1.1.3 Term lengths for members of the popular house shall reflect the need for accountability through regular and periodic legislative elections.

  1.2 Candidate Eligibility

    1.2.1 Restrictions on candidate eligibility shall not be based on religion, gender, ethnicity, race, or disability.

    1.2.2 Special measures to encourage the political participation of marginalized groups shall be narrowly drawn to accomplish precisely defined, and time-limited, objectives.

  1.3 Incompatibility of Office

    1.3.1 No elected member shall be required to take a religious oath against his or her conscience in order to take his or her seat in the legislature.

    1.3.2 In a bicameral legislature, a legislator may not be a member of both houses.

    1.3.3 A legislator may not simultaneously serve in the judicial branch or as a civil servant of the executive branch.

1.4 Immunity

1.4.1 Legislators shall have immunity for anything said in the course of the proceedings of legislature.

1.4.2 Parliamentary immunity shall not extend beyond the term of office, but a former legislator shall continue to enjoy protection for his or her term of office.

1.4.3 The executive branch shall have no right or power to lift the immunity of a legislator.

1.4.4 Legislators must be able to carry out their legislative and constitutional functions in accordance with the constitution, free from interference.

1.5 Remuneration and Benefits

1.5.1 The legislature shall provide proper remuneration and reimbursement of parliamentary expenses to legislators for their service, and all forms of compensation shall be allocated on a nonpartisan basis.

1.6 Resignation

1.6.1 Legislators shall have the right to resign their seats.

1.7 Infrastructure

1.7.1 The legislature shall have adequate physical infrastructure to enable members and staff to fulfill their responsibilities.

II. Organization of the Legislature

2. Procedure and Sessions

2.1 Rules of Procedure

2.1.1 Only the legislature may adopt and amend its rules of procedure.

2.2 Presiding Officers

2.2.1 The legislature shall select or elect presiding officers pursuant to criteria and procedures clearly defined in the rules of procedure.

2.3 Convening Sessions

2.3.1 The legislature shall meet regularly, at intervals sufficient to fulfill its responsibilities.

2.3.2 The legislature shall have procedures for calling itself into regular session.

2.3.3 The legislature shall have procedures for calling itself into extraordinary or special session.

2.3.4 Provisions for the executive branch to convene a special session of the legislature shall be clearly specified.

2.4 Agenda

2.4.1 Legislators shall have the right to vote to amend the proposed agenda for debate.

2.4.2 Legislators in the lower or only house shall have the right to initiate legislation and to offer amendments to proposed legislation.

2.4.3 The legislature shall give legislators adequate advance notice of session meetings and the agenda for the meeting.

2.5 Debate

2.5.1 The legislature shall establish and follow clear procedures for structuring debate and determining the order of precedence of motions tabled by members.

2.5.2 The legislature shall provide adequate opportunity for legislators to debate bills prior to a vote.

2.6 Voting

2.6.1 Plenary votes in the legislature shall be public.[1]

2.6.2 Members in a minority on a vote shall be able to demand a recorded vote.

2.6.3 Only legislators may vote on issues before the legislature.

2.7 Records

2.7.1 The legislature shall maintain and publish readily accessible records of its proceedings.

3. Committees

3.1 Organization

3.1.1 The legislature shall have the right to form permanent and temporary committees.

3.1.2 The legislature's assignment of committee members on each committee shall include both majority and minority party members and reflect the political composition of the legislature.

3.1.3 The legislature shall establish and follow a transparent method for selecting or electing the chairs of committees.

3.1.4 Committee hearings shall be in public. Any exceptions shall be clearly defined and provided for in the rules of procedure.

3.1.5 Votes of committee shall be in public. Any exceptions shall be clearly defined and provided for in the rules of procedure.

3.2 Powers

3.2.1 There shall be a presumption that the legislature will refer legislation to a committee, and any exceptions must be transparent, narrowly defined, and extraordinary in nature.

3.2.2 Committees shall scrutinize legislation referred to them and have the power to recommend amendments or amend the legislation.

3.2.3 Committees shall have the right to consult and/or employ experts.

3.2.4 Committees shall have the power to summon persons, papers, and records, and this power shall extend to witnesses and evidence from the executive branch, including officials.

3.2.5 Only legislators appointed to the committee, or authorized substitutes, shall have the right to vote in committee.

3.2.6 Legislation shall protect informants and witnesses presenting relevant information to commissions of inquiry about corruption or unlawful activity.

4. Political Parties, Party Groups, and Cross-Party Groups

    4.1 Political Parties

        4.1.1 The right of freedom of association shall exist for legislators, as for all people.

        4.1.2 Any restrictions on the legality of political parties shall be narrowly drawn in law and shall be consistent with the International Covenant on Civil and Political Rights.

    4.2 Party Groups

        4.2.1 Criteria for the formation of parliamentary party groups, and their rights and responsibilities in the legislature, shall be clearly stated in the Rules.

        4.2.2 The legislature shall provide adequate resources and facilities for party groups pursuant to a clear and transparent formula that does not unduly advantage the majority party.[2]

    4.3 Cross-Party Groups

        4.3.1 Legislators shall have the right to form interest caucuses around issues of common concern.

5. Parliamentary Staff

    5.1 General

        5.1.1 The legislature shall have an adequate nonpartisan professional staff to support its operations, including the operations of its committees.

        5.1.2 The legislature, rather than the executive branch, shall control the parliamentary service and determine the terms of employment.

        5.1.3 The legislature shall draw and maintain a clear distinction between partisan and nonpartisan staff.

        5.1.4 Members and staff of the legislature shall have access to sufficient research, library, and information, communication, and technology facilities.

    5.2 Recruitment

        5.2.1 The legislature shall have adequate resources to recruit staff sufficient to fulfill its responsibilities. The rates of pay shall be broadly comparable to those in the public service.

        5.2.2 The legislature shall not discriminate in its recruitment of staff on the basis of race, ethnicity, religion, gender, disability, or, in the case of nonpartisan staff, party affiliation.

    5.3 Promotion

        5.3.1 Recruitment and promotion of nonpartisan staff shall be on the basis of merit and equal opportunity.[3]

    5.4 Organization and Management[4]

        5.4.1 The head of the parliamentary service shall have a form of protected status to prevent undue political pressure.

        5.4.2 Legislatures should, either by legislation or resolution, establish corporate bodies responsible for providing services and funding entitlements for parliamentary purposes and providing for governance of the parliamentary service.

5.4.3 All staff shall be subject to a code of conduct.

III. Functions of the Legislature

6. Legislative Function

6.1 General

6.1.1 The approval of the legislature is required for the passage of all legislation, including budgets.

6.1.2 Only the legislature shall be empowered to determine and approve the budget of the legislature.

6.1.3 The legislature shall have the power to enact resolutions or other nonbinding expressions of its will.

6.1.4 In bicameral systems, only a popularly elected house shall have the power to bring down government.

6.1.5 A chamber where a majority of members are not directly or indirectly elected may not indefinitely deny or reject a money bill.

6.2 Legislative Procedure

6.2.1 In a bicameral legislature, there shall be clearly defined roles for each chamber in the passage of legislation.

6.2.2 The legislature shall have the right to override an executive veto.

6.3 The Public and Legislation

6.3.1 Opportunities shall be given for public input into the legislative process.

6.3.2 Information shall be provided to the public in a timely manner regarding matters under consideration by the legislature.

7. Oversight Function

7.1 General

7.1.1 The legislature shall have mechanisms to obtain information from the executive branch sufficient to exercise its oversight function in a meaningful way.

7.1.2 The oversight authority of the legislature shall include meaningful oversight of the military security and intelligence services.

7.1.3 The oversight authority of the legislature shall include meaningful oversight of state-owned enterprises.

7.2 Financial and Budget Oversight

7.2.1 The legislature shall have a reasonable period of time in which to review the proposed national budget.[5]

7.2.2 Oversight committees shall provide meaningful opportunities for minority or opposition parties to engage in effective oversight of government expenditures. Typically, the public accounts committee will be chaired by a member of the opposition party.

7.2.3 Oversight committees shall have access to records of executive branch accounts and related documentation sufficient to be able to meaningfully review the accuracy of executive branch reporting on its revenues and expenditures.

7.2.4 There shall be an independent, nonpartisan supreme or national audit office whose reports are tabled in the legislature in a timely manner.

7.2.5 The supreme or national audit office shall be provided with adequate resources and legal authority to conduct audits in a timely manner.

7.3 No Confidence and Impeachment

7.3.1 The legislature shall have mechanisms to impeach or censure officials of the executive branch or express no confidence in the government.

7.3.2 If the legislature expresses no confidence in the government, the government is obliged to offer its resignation. If the head of state agrees that no other alternative government can be formed, a general election should be held.

8. Representational Function

8.1 Constituent Relations

8.1.1 The legislature shall provide all legislators with adequate and appropriate resources to enable the legislators to fulfill their constituency responsibilities.

8.2 Parliamentary Networking and Diplomacy

8.2.1 The legislature shall have the right to receive development assistance to strengthen the institution of parliament.

8.2.2 Members and staff of parliament shall have the right to receive technical and advisory assistance, as well as to network and exchange experience with individuals from other legislatures.

IV. Values of the Legislature

9. Accessibility

9.1 Citizens and the Press

9.1.1 The legislature shall be accessible and open to citizens and the media, subject only to demonstrable public safety and work requirements.

9.1.2 The legislature should ensure that the media are given appropriate access to the proceedings of the legislature without compromising the proper functioning of the legislature and its rules of procedure.

9.1.3 The legislature shall have a nonpartisan media relations facility.

9.1.4 The legislature shall promote the public's understanding of the work of the legislature.

9.2 Languages

9.2.1 Where the constitution or parliamentary rules provide for the use of multiple working languages, the legislature shall make every reasonable effort to provide for simultaneous interpretation of debates and translation of records.

10. Ethical Governance

    10.1 Transparency and Integrity

        10.1.1 Legislators should maintain high standards of accountability, transparency, and responsibility in the conduct of all public and parliamentary matters.

        10.1.2 The legislature shall approve and enforce a code of conduct, including rules on conflicts of interest and the acceptance of gifts.

        10.1.3 Legislatures shall require legislators to fully and publicly disclose their financial assets and business interests.

        10.1.4 There shall be mechanisms to prevent, detect, and bring to justice legislators and staff engaged in corrupt practices.

Annex 3B: Recommended Benchmarks for Codes of Conduct for Parliamentarians

Purpose and Role of Parliamentary Code of Conduct

1.1 A Code of Conduct forms an important part of the Integrity System.[6]

1.2 Codes affecting the conduct of individual Members of Parliament encourage ethical conduct, reduce risks to the integrity of the Parliament as the paramount political institution, enable it to perform its functions more effectively, enhance propriety, and strengthen the community's trust in Parliament.

1.3 A Code of Conduct for Members of Parliament applies to all office holders who are members of the House of Parliament, including the Presiding Officer, the Prime Minister/Premier/Chief Minister and the ministers.

1.4 A Code of Conduct includes both aspirational provisions (what parliamentarians ought to do) and prescriptive provisions (what parliamentarians must do or not do) and should be seen as the minimum standard for conduct.[7]

1.5 Codes of Conduct have a purpose different from Standing Orders, which are primarily rules of procedure.

1.6 A code should be written in a style that is simple, clear, and specific.

Principles

2.1 A Member of Parliament as a public officer exercises a public trust.[8]

2.2 Members of Parliament shall behave according to the following principles:

- **Selflessness.** Members of Parliament should act solely in terms of the public interest.
- **Integrity.** Members of Parliament must avoid placing themselves under any obligation to people or organizations that might try inappropriately to influence them in their work. They should not act or take decisions in order to gain financial or other material benefits for themselves, their family, or their friends. They must declare and resolve any interests and relationships.

- *Objectivity.* Members of Parliament must act and take decisions impartially, fairly and on merit, using the best evidence and without discrimination or bias.
- *Accountability.* Members of Parliament are accountable to the public for their decisions and actions and must submit themselves to the scrutiny necessary to ensure this.
- *Openness.* Members of Parliament should act and take decisions in an open and transparent manner. Information should not be withheld from the public unless there are clear and lawful reasons for so doing.
- *Honesty.* Members of Parliament should be truthful.
- *Leadership.* Members of Parliament should exhibit these principles in their own behaviour. They should actively promote and robustly support the principles and be willing to challenge poor behaviour wherever it occurs.[9,10]

2.3   Members of Parliament shall:
- Act in good conscience
- Respect the intrinsic dignity of all
- Act so as to merit the trust and respect of the community
- Give effect to the ideals of democratic government and abide by the letter and spirit of the Constitution and uphold the separation of powers and the rule of law
- Hold themselves accountable for conduct for which they are responsible
- Exercise the privileges and discharge the duties of public office diligently and with civility, dignity, care and honour.[11]

2.4   Members of Parliament have individual responsibility as contributors to the functioning of the institution.

2.5   Parliamentary immunity (i.e. parliamentary privilege) protects the right of Members of Parliament to speak in parliament without fear of prosecution or suit for defamation.

2.6   Members of Parliament shall respect the roles, independence, rights and responsibilities of parliamentary staff.

2.7   In a parliamentary democracy, every Member of Parliament has a responsibility to ensure that the Executive Government is accountable to the Parliament.

Benchmarks for Codes of Conduct for Parliamentarians

3.1   **Disclosure and publication of interests.** The code shall indicate that each Member shall disclose every interest which may create a perception of conflict between an interest and the duties and responsibilities set out in PRINCIPLES. It shall prescribe provisions to which each Member is subject, with provisions to the effect as follows.

3.1.1   Each Member shall disclose to the Parliament all relevant interests that a reasonable person might think could give rise to the perception of influencing behaviour between the Member's duties and responsibilities and his/her personal interests (e.g. land and property assets, share-holdings, gifts,[12] foreign travel, symbolic rewards (e.g. honorary degree), sources of income, remunerated employment, directorships, liabilities, hospitality and affiliations). These may be subject to specified thresholds. This applies to items received and could also apply to items donated or given. These shall be disclosed immediately following election and continuously updated within a reasonable period specified by the parliament above a specified threshold.

3.1.2 A Member shall not vote in a division on a question about a matter, other than public policy (i.e. government policy, not identifying any particular person individually and immediately) in which he or she has a particular direct pecuniary interest above a threshold (if specified).[13]

3.1.3 A Member shall not use for personal benefit confidential information (i.e. non-public information) gained as a public officer.

3.1.4 There should be an effective mechanism to verify any disclosure and to immediately notify any discrepancy in a public report to the House.

3.1.5 The Parliament shall publish the interests disclosed and the purposes and amounts of expenditure of public funds by each Member as soon as practicable in the most accessible means available (e.g. parliamentary website).

3.1.6 These provisions also apply to interests held by the member's spouse or close family members.

3.2 Use of public property. A code should make provision to the effect that a Member may use public funds, property or facilities only in the public interest and as permitted by law (does not include for political party purposes).

3.3 Inducements

3.3.1 A Member shall not accept any form of inducement that could give rise to conflict of interest or influence behaviour.

3.3.2 A member shall not engage in paid lobbying, paid parliamentary advice or paid advocacy.

3.3.3 A Member shall not use his or her position to seek or secure future employment, paid lobbying, consultancy work or other remuneration or benefit upon ceasing to be a Member of Parliament.

3.3.4 A Member shall represent the interests of constituents on an equitable basis and not on the basis of personal or political affiliations, or inducements.

3.4 Civility

Members shall treat each other, the Parliament and the people with respect, dignity and courtesy, including parliamentary staff.

3.5 Behaviour

A Member shall not assault, harass, or intimidate another person.

3.6 Attendance

Every member shall attend every sitting of the House, in accordance with practice of the House, except with reasonable excuse, or in the case of extended absences, if excused in accordance with the practice of the House.

Ethics Adviser

As part of an effective implementation of a Code of Conduct, advice shall be available to individual MPs to help them decide how to deal with ethical dilemmas. A code of conduct may provide for an ethics adviser according to the following model.

4.1 The adviser shall be independent of influence by any person in giving advice. (The House should designate the title of the office.[14])

4.2 The adviser shall be selected by a non-partisan process or other method designed to secure multiparty support.

4.3 The adviser shall have knowledge, experience, personal qualities and standing within the community suitable to the office; skill in professional ethics or law is desirable.

4.4 The Code shall protect the adviser from removal except for proven misbehaviour or other reasonable grounds.

4.5 Members shall endeavour to routinely discuss ethical dilemmas with an ethics adviser.

4.6 Members who, if unable to discuss an ethical dilemma with an ethics adviser or having done so, remain in doubt, must act with caution and not engage in any potentially compromising action.

4.7 Advice may be sought on conflicts of interest and any issue arising from codes of conduct and ethics and integrity issues.

4.8 The adviser shall base advice in each instance on the facts as related by the MP and any other relevant facts of which s/he becomes aware.

4.9 The adviser shall not disclose the fact that s/he has been consulted, nor any information provided by the MP or any advice given to the MP.

4.10 Advice sought and given is confidential, and shall not be accessible through provisions for freedom of information. However, the person who seeks written advice may make it, and the related request, public.

4.11 The adviser shall not investigate any complaint.

Enforcement

As part of the effective implementation of a code, an independent system for investigating alleged breaches should be established. A suggested model follows:

5.1 **Complaints and Investigations.** A code shall make provisions to the effect that:

5.1.1 A complaint alleging breach of the Code by a Member shall be made to an identified office holder, who must forthwith refer it to an investigator for investigation of the facts.

5.1.2 At least one investigator must be appointed by the House as soon as practicable following adoption of the Code.

5.1.3 An Investigator shall be independent of Parliament, any Member of the Parliament, Government, or political party or grouping, and is appointed for a fixed term.

5.1.4   The investigator must be selected by a non-partisan process or other method designed to secure multiparty support.

5.1.5   An Investigator shall have knowledge, investigative skills, experience, personal qualities and standing within the community suitable to the office.

5.1.6   The Code shall protect the investigator from removal except for proven misbehaviour or other reasonable grounds.

5.1.7   The investigator may determine that a complaint is frivolous or vexatious and decline to investigate it.

5.1.8   A Member and the complainant shall treat any complaint as if *sub judice*.

5.1.9   Any Member of Parliament shall cooperate with and assist an Investigator in the investigation of any complaint under the Code.

5.1.10 If there is evidence of a breach of criminal law, it must forthwith be referred to the police or corruption control agency as appropriate.

5.1.11 After investigation, the investigator must present a report to the Presiding Officer (or Deputy if concerning the Presiding Officer), who must determine whether or not a breach has occurred, and if a breach has occurred, refer the report to the House for further proceedings in accordance with its rules.

5.1.12 If a complaint has become known publicly and has not been upheld, this outcome shall be made public.

5.2 Appeal or review. The Code shall make provision that a Member against whom a complaint has been upheld, has rights to appeal or review.

5.3 Sanctions and penalties

5.3.1   The Code shall specify graduated sanctions and penalties for breaches of the Code according to the seriousness of the effects of breaches on the functioning, reputation and legitimacy of the parliament.

5.3.2   The Code shall specify that a Member convicted of a breach of the criminal law, may in addition be subject to a sanction or penalty if found to have breached the Code.

Making and Updating the Code

6.1 The House shall ensure that its Code of Conduct remains relevant, is reviewed and revised periodically, is up to date and is familiar to its Members of Parliament.

6.1.1   The Code shall be made by the House of Parliament, whose Members are to be subject to its provisions (i.e. by each House in a bicameral Parliament) and remains in force unless and until remade.

6.1.2   The Code shall be established by a decision of the House of Parliament to which it relates.

6.1.3   The Code shall be subject to continuous and regular review. A mechanism shall be established for this purpose and to report to the parliament on its operation immediately following each general election, in response to requests by the Presiding Officer, and at such other times as it wishes.

Fostering a Culture of Ethical Conduct

7.1 Each House should sustain a culture of ethical conduct reflecting a sound understanding of the parliamentary role, the public interest and the institution of parliament. Such a culture may be facilitated by:

7.1.1 Introductory and continuing education to assist Members to enhance their skills in ethical deliberation.

7.1.2 Induction, which includes mentoring and experience-sharing activities involving both new and experienced Members.

7.1.3 Exemplary behaviour by those in leadership roles.

7.1.4 Endeavours to detect and act to deter even minor breaches from which serious breaches may develop.

7.1.5 Members being encouraged to consult with the Ethics Advisor before acting on a matter that raises ethical issues.

7.1.6 Members acknowledging and accepting provisions of a Code of Conduct when swearing an Oath or making an Affirmation.

7.1.7 Publishing and making available the Code to both Members and the public.

7.1.8 Ensuring that newly elected members receive induction in the Code of Conduct, and engaging in self-assessment of their individual ethical competence.

7.1.9 Encouraging discussions with the ethics adviser, which shall be treated as routine and normal, with frequent informal contact between the ethics adviser and the Members.

7.1.10 Requiring every Member to participate in activities to enhance their ethical competence on a regular basis. These activities could be online, if resources permit.

7.1.11 Requiring Members to provide evidence on a regular basis that they have read and understood the provisions of the Code.

7.1.12 Endeavouring to adapt the code to changing expectations of society with regard to ethical conduct.

Notes

1. The study group noted that one possible exception to this provision might be the election of officers.

2. The study group considered it best practice for legislatures to provide party groups with funding allocations and to allow each party group to make its own decisions on the types of facilities it required. The study group recognized the special circumstances of small and underresourced jurisdictions.

3. Rather than banning political activity by nonpartisan staff members, the study group recommended that all staff members be subject to a code of conduct and

assessed on their conduct annually. A code of conduct should make clear what constitutes acceptable staff behavior and serve to prevent staff members from using their positions to influence the functioning of the legislature in a political manner.

4. Benchmarks 5.4.1 and 5.4.2 were taken directly from the recommendations of the previous CPA's study group meeting on the Financing and Administration of Parliament, which was held in Zanzibar, Tanzania, on May 25–29, 2005.

5. The study group made reference to the Organisation for Economic Co-operation and Development's best-practice guidelines, which suggest presentation of the draft budget to the legislature no less than three months prior to the start of the fiscal year (OECD 2002).

6. Integrity Systems are a form of risk management that provide insurance against corruption. They include norms (e.g. ethical behaviour), institutions (e.g. corruption control commission), and mechanisms (e.g. special investigative powers) designed to reduce corruption and enhance integrity. The extent, strength, and degree of interconnectedness (including systemic and non-systemic dimensions), overlaps, conflicts and mutual supports affect how an integrity system actually works (Sampford 2014).

7. The Australian House of Representatives Committee reported that codes of conduct it had examined seemed to fall into the two categories: prescriptive or aspirational. One approach is to establish a more directive or prescriptive code, which would include quite detailed rules and be a rather lengthy statement. The aim of a prescriptive code is to provide a comprehensive account of the conduct required of members in all conceivable situations:

> The alternative approach is for a more aspirational set of principles from which each member must determine his or her own behaviour. An aspirational code aims to provide a frame of reference for making decisions that involve competing values ((House of Representatives Standing Committee of Privileges and Members' Interests (Australia), 2011), p. 29).

Few if any codes are solely either aspirational or prescriptive. A code including both aspirational and prescriptive provisions is more likely to be effective according to the research leading to these Benchmarks.

8. As a holder of public office, a Member must avoid:

- official misconduct that involves a breach of powers and duties entrusted to a Member for the public benefit and in which the Member has abused them or his position;
- wilful neglect of duty;
- wilfully embarking on a course of action which the Member has no legal right to undertake;
- oppression and extortion;
- incompatible positions;
- arrangements which are in conflict with his or her official duties;
- bribery;
- misuse of public property.

[adapted from Smith (2014)]

9. These principles are adapted from "The Seven Principles of Public Life" (the "Nolan Principles") for holders of public office (Nolan 1995).

10. See also the general principles to govern the conduct of members of relevant authorities in England and police authorities in Wales as follows:

Selflessness

i.   Members should serve only the public interest and should never improperly confer an advantage or disadvantage on any person.

Honesty and Integrity

ii. Members should not place themselves in situations where their honesty and integrity may be questioned, should not behave improperly and should on all occasions avoid the appearance of such behaviour.

Objectivity

iii. Members should make decisions on merit, including when making appointments, awarding contracts, or recommending individuals for rewards or benefits.

Accountability

iv. Members should be accountable to the public for their actions and the manner in which they carry out their responsibilities, and should co-operate fully and honestly with any scrutiny appropriate to their particular office.

Openness

v.   Members should be as open as possible about their actions and those of their authority, and should be prepared to give reasons for those actions.

Personal Judgement

vi. Members may take account of the views of others, including their political groups, but should reach their own conclusions on the issues before them and act in accordance with those conclusions.

Respect for Others

vii. Members should promote equality by not discriminating unlawfully against any person, and by treating people with respect, regardless of their race, age, religion, gender, sexual orientation or disability. They should respect the impartiality and integrity of the authority's statutory officers, and its other employees.

Duty to Uphold the Law

viii. Members should uphold the law and, on all occasions, act in accordance with the trust that the public is entitled to place in them.

Stewardship

ix. Members should do whatever they are able to do to ensure that their authorities use their resources prudently and in accordance with the law.

Leadership

x.   Members should promote and support these principles by leadership, and by example, and should act in a way that secures or preserves public confidence.

[Statutory Instrument 2001 No. 1401. The **Relevant Authorities (General Principles) Order 2001** (United Kingdom). Retrieved 18 March 2015 from http://www.tisonline .net/ContentUploads/CaseUploads/RelAuthOrder_6102009154823.doc]

11. This section is adapted from the "Politicians' Pledge" (St. James Ethics Centre 2015).

12. This is not to suggest a total ban on accepting or donating gifts, but it recognizes that the very act of offering or receiving a gift establishes a favorable predisposition to the

other person, irrespective of the value of the gift (Malmendier and Schmidt 2012). Total bans on accepting any gifts risk leading to failure by even the most ethical Members of Parliament. Once a person is tainted as unethical for accepting or offering a gift, no matter how commonplace, reasonable, and harmless a social behaviour, critics have a tool with which to tar and tarnish the reputation of the individual and other Members of Parliament (Kania 2004). Disclosure greatly reduces the risk of appearance of impropriety.

13. Adapted from House of Representatives Practice (House of Representatives [Australia] 2012).

14. Examples of titles include: Conflict of Interest and Ethics Commissioner; Parliamentary Ethics Adviser; (Parliamentary) Integrity Commissioner; Parliamentary Commissioner for Standards.

References

CPA (Commonwealth Parliamentary Association). 2006. "Recommended Benchmarks for Democratic Legislatures." CPA London. http://wbi.worldbank.org/wbi/Data/wbi/wbicms/files/drupal-acquia/wbi/Recommended%20Benchmarks%20for%20Democratic%20Legislatures.pdf.

———. 2009. "Recommended Benchmarks for Pacific Island Democratic Legislatures." CPA London. http://www.cpahq.org/cpahq/cpadocs/6_Recommended%20Benchmarks%20for%20Pacific%20Island%20Democratic%20Legislatures.pdf.

———. 2010. "Recommended Benchmarks for Asia, India, and South-East Asia Regions' Democratic Legislatures." CPA London. http://www.agora-parl.org/sites/default/files/CPA%20Recommended%20Benchmarks%20for%20Asia%2C%20India%20and%20South%20East%20Asia%20Regions%20Democratic%20Legislatures.pdf.

———. 2011. "Recommended Benchmarks for the CPA Caribbean, Americas, and Atlantic Region Democratic Legislatures." CPA London. http://www.cpa-caaregion.org/media/get_media.php?mediaid=caa4fafb-a31.

CPA (Commonwealth Parliamentary Association), Commonwealth Legal Education Association, Commonwealth Magistrates' and Judges' Association, and Commonwealth Lawyers' Association. 2004. "Commonwealth (Latimer House) Principles on the Three Branches of Government." Commonwealth Secretariat, London.

House of Representatives [Australia]. 2012. House of Representatives Practice (6th ed.). Canberra: Department of the House of Representatives.

Kania, Richard R. E. 2004. "The Ethical Acceptability of Gratuities: Still Saying 'Yes' after All These Years." *Criminal Justice Ethics* 23 (1): 54–63.

Malmendier, Ulrike, and Klaus Schmidt. 2012. "You Owe Me." NBER Working Paper 18453, National Bureau of Economic Research, Cambridge, MA.

Nolan, Lord Michael. 1995. "Standards in Public Life: First Report of the Committee on Standards in Public Life." Presented to Parliament by the Prime Minister by Command of Her Majesty, May 1995. CM 2850-I.

OECD (Organisation for Economic Co-operation and Development). 2002. "OECD Best Practices for Budget Transparency." OECD, Paris.

SADC PF (Southern African Development Community Parliamentary Forum). 2010. "Benchmarks for Democratic Legislatures in Southern Africa." SADC PF, Windhoek. http://www.agora-parl.org/node/2777.

Sampford, Charles. 2014. "Integrity Systems: Some History." Paper presented at the "Assessing National Integrity Systems in the G20 and Beyond" workshop. Transparency International, Brisbane.

Smith, Tim. 2014. "Integrity in Politics? Public Office as Public Trust? Is There Hope?" Paper presented to the University of the Third Age, Hawthorne.

St. James Ethics Centre. 2015. "Politicians' Pledge." St. James Ethics Centre, Sydney, Australia.

CHAPTER 4

The Democratic Reality of Parliaments: What Evaluation Criteria?

M. Pascal Terrasse

Introduction

Over the years, interest has grown in the use of indicators to assess and improve the functioning of parliaments. Only very recently, however, have the key organizations in the field of parliamentary development begun to wonder about the criteria of parliamentary democracy. This reflection marked the beginning of a structured process to develop benchmarks and self-assessment tools for democratic parliaments. Indeed, several interparliamentary associations, such as the Inter-Parliamentary Union and the Commonwealth Parliamentary Association, have since published operating guidelines for parliamentary best practices and developed a compendium of benchmarks for democratic parliaments.

Because existing benchmarking tools more closely reflect the needs of parliaments in the Anglo-Saxon world, an additional set of tools was needed for francophone parliaments. Hence, the Assemblée Parlementaire de la Francophonie (Parliamentary Assembly of La Francophonie, or APF), at the instigation of its former secretary general, Jacques Legendre, in July 2008 joined international discussions on the criteria of parliamentary democracy in a search to emphasize the specificities of the francophone world. To do so, the APF committed, in collaboration with the United Nations Development Programme (UNDP), to produce this chapter as a reference document for stakeholders involved in international discussions on parliamentary democracy.[1] In particular, the aim of this chapter is to present a collection of progressive criteria and objectives for APF parliamentary members to strive for.

To better understand the importance of evaluation criteria, the next section of this chapter provides an overview of the APF and its role in the francophone world. The following section then discusses the process of developing criteria for evaluation. The subsequent section summarizes the different criteria by category.

This summary is followed by a discussion of the challenges of the approach for francophone countries. Finally, the last section concludes with reference to benefits of the approach and next steps.

Role of the APF

The APF is the advisory assembly of parliaments from francophone countries. It brings together 81 parliaments and parliamentary associations from every continent. It creates a venue for debates, proposals, and exchanges of information on all topics that are of interest and concern to its members and constituents. Through its counsels and recommendations to the Ministerial Conference of La Francophonie and the Permanent Council of La Francophonie, it participates in the institutional life of francophone countries. It also intervenes before heads of state and governments during the Summit of La Francophonie. Among its objectives, the APF supports the promotion of democracy, peace, cultural diversity, and human rights; the spread of the French language; and the valuable role of parliamentarians in a democratic society.

The assembly consists of parliaments from four geographic branches: Africa, the Americas, Asia-Pacific, and Europe. Parliamentarians from each branch meet at annual plenary sessions. APF activities take place through four standing committees: the Committee on Parliamentary Affairs; the Political Committee; the Committee on Development and Cooperation; and the Committee on Education, Cultural Affairs, and Communication. Moreover, the APF has the Network of Women Parliamentarians for female parliamentarians from francophone countries to work together on topics that concern them more specifically. Finally, within the Committee on Education, Cultural Affairs, and Communication is a structure called the Parliamentary Network for the Fight against HIV/AIDS, which facilitates the APF's participation in discussions on the fight against this pandemic. Meetings of these committees and networks are held twice a year, either during the annual plenary session or the respective intercessional meeting. In addition to annual conferences, parliamentarians from every branch convene annually at regional meetings.

An important focus of the APF's work is implementing actions in the areas of interparliamentary cooperation and democracy development in close collaboration with the Organisation Internationale de la Francophonie (International Organization of La Francophonie, or OIF). The aim of these actions is to strengthen solidarity among parliamentary institutions and promote democracy and the rule of law, especially in the francophone world. The APF is often called the watchdog of democracy because of its democratic standards. Indeed, in accordance with the fundamental principles underlying the assembly, if the constitutional order of an APF member government has been overthrown and the parliament of that country is dissolved or deprived of its powers, then the APF branch representing the parliament is suspended until constitutional order is restored.

The APF plays a very active role in improving the operation and working methods of parliaments in francophone countries. To this end, it implements a number of parliamentary cooperation workshops, such as the following:

- A large-scale parliamentary information management program, the Noria project, which takes into account the specific needs of participating parliaments
- Training seminars for parliamentarians
- Training workshops for senior officials in the parliamentary branches
- The Francophone Youth Parliament, which familiarizes youth with the functioning of democratic institutions
- Participation of the APF's parliamentarians in electoral observation missions in francophone countries

The APF is also involved in coordinating the Initiative Multilatérale de Coopération Interparlementaire Francophone (Multilateral Initiative for Francophone Interparliamentary Cooperation, or IMCIF), an innovative cooperation mechanism for the parliaments of French-speaking states emerging from crisis or in the process of democratic consolidation. The IMCIF aims to unite international support for such parliaments as a way to streamline and harmonize actions taken on their behalf while promoting effective, results-oriented cooperation between francophone parliaments. Since 2012, the National Assembly of Côte d'Ivoire benefits from an IMCIF capacity-building scheme. Since 2014, this program has been extended to the National Assemblies of Mali and Madagascar under the name "Programme multilateral de développement parlementaire francophone."

Development Process of the Evaluation Method

The APF aims to play an active role in global thinking on developing and strengthening parliamentary democracy. To this end, it has developed criteria for evaluating the democratic reality of parliaments for the francophone world. This section discusses the process of developing these criteria.

To first launch this project, the APF established a close collaboration with UNDP through several measures. In particular, in July 2008, the APF shared a copy of an unpublished UNDP report by Jean-Philippe Roy, a lecturer in political science at the Université François-Rabelais, with each of its parliamentary branches. Titled "Democratization of Parliaments," the study coincided with the signing of a cooperation agreement between the APF and UNDP in New York in January 2009.

Also in 2008, an in-depth review process was begun to produce a reference document on evaluation criteria for democratic parliaments. First, with the support of UNDP and the National Assembly of Quebec, an intern from Laval University joined the APF secretary general in October 2008 to prepare a rough draft of this reference document in the form of a comparative study of the standing orders of several parliaments in the francophone world.[2] Subsequently, a preliminary draft of the reference document underwent a critical review by the APF's different branches at its New York office in January 2009 and by the Network of Women Parliamentarians in February 2009. Soon thereafter, francophone parliamentarians, academics, and representatives of the Association of Secretaries General of Francophone Parliaments held debates on and proposed amendments to the draft document at the Committee on Parliamentary Affairs (Switzerland, March 23–25) and the Committee on Political Affairs (the Lao People's Democratic Republic, April 9–11).

The APF secretary general then forwarded a revised draft incorporating proposed amendments to the APF branches for a last round of comments and produced a consolidated draft document. Final changes to the document were made by the APF's executive, the document was endorsed by the two committees involved, and then it was finally adopted at the 35th session of the APF in Paris in July 2009.

In October 2009, the reference document was delivered at the International Conference on Best Practices for Parliamentary Democracy, which was held in Geneva. Organized by UNDP and the World Bank, this meeting aimed at consolidating the work undertaken by the APF, the Commonwealth Parliamentary Association, the World Bank, the Inter-Parliamentary Union, and the National Democratic Institute. It was also presented by Pierre de Bané, a Canadian senator and chair of the Parliamentary Affairs Committee of the APF, at the International Conference on Benchmarking and Self-Assessment for Democratic Parliaments, sponsored by the World Bank and UNDP and held in Paris in March 2010.

Specific Content of the Evaluation Criteria

In recent years, a strong trend has favored increasing democratic reforms. Parliaments are striving to be more open and transparent in their procedures; more independent from the government; and more representative, accessible, and responsive to their constituents through the use of new technological tools. Moreover, a large number of parliaments have adopted funding rules for political parties and ethical codes to restore public confidence in the integrity of parliamentarians. A number of parliaments have examined measures aimed at ensuring better political and sociological representation, such as provisions to increase the representation of women and minorities.

Democracies must adapt continually in an ever-changing world. In such an environment, only exchange and synergy can stimulate and optimize parliamentary reflections on how to develop and strengthen democracy. The APF's aim in developing criteria is to strengthen the capacity of parliaments to conduct their own self-assessments on the basis of standards designed in accordance with democratic norms. Moreover, donors, organizations, and institutions that offer parliamentary strengthening programs can use these criteria to develop more responsive support programs and to establish well-defined guidelines for their assistance.

Because the APF unites parliaments from different countries, it aims at recognizing the contributions of the diverse parliamentary traditions in the francophone world. These traditions reflect, each in their own way, the wealth of the organization. The APF's challenge in developing evaluation criteria was to synthesize these differences in a single document. In doing so, the APF developed a set of criteria that can be categorized into four major components: (a) elections and the status of parliamentarians, (b) prerogatives of parliament, (c) organization of parliament, and (d) parliament and communication. This section reviews each of these components in detail.

Elections and the Status of Parliamentarians

The first component of the evaluation criteria relates to elections and the status of parliamentarians. With respect to elections, the national constitution must reflect an understanding of the basic rules of the electoral process and the status of parliamentarians. Parliamentarians must be elected by universal suffrage through free, reliable, and transparent elections in accordance with national and international standards. However, second chambers can be governed by specific rules provided in the constitution or by national legislation.

The evaluation criteria emphasize that elections must be held at regular intervals, and the legislature must be limited in duration and hold new elections before expiration. Similarly, elections must take place without any interference or attack on liberty, physical integrity, freedom of opinion and expression, freedom to organize meetings and demonstrations, and freedom of association by any citizen or candidate.

From the initial operations and election campaign to counting the votes and proclaiming the results, the preparation and management of elections must be assigned to entities with powers to perform a rigorous electoral monitoring process. This assignment is key to ensuring the trustworthiness of the ballot, the full participation of citizens in the electoral process, and equal treatment of candidates throughout the electoral process. Consequently, the legally constituted political parties must be able to participate in all stages of the electoral process, in accordance with the democratic principles enshrined in the basic texts and primary institutions. Furthermore, an independent and impartial judicial authority is needed to manage any electoral disputes that may arise.

Finally, a candidate should not be disqualified because of gender, race, language, religion, economic situation, physical disability, or considerations of respect for his or her privacy. However, specific procedures may provide for the representation of national and regional diversity and its components.

In terms of the status of parliamentarians, several criteria provide guidance for APF member parliaments. For instance, in view of possible incompatibilities within a parliament, the criteria established that an elected official cannot be required to submit to a religious oath against his or her conscience. Moreover, in a bicameral parliament, a parliamentarian cannot simultaneously be a member of both houses. In addition, parliamentary incompatibilities must be defined by legislation and monitoring, and their sanctions should be the subject of a special procedure.

With respect to immunity and parliamentary privileges, all parliamentary members must be able to fulfill their mandate freely and be sheltered from any undue pressure or influence. Parliamentarians cannot be pursued, wanted, arrested, detained, tried, or imprisoned because of their votes or their views expressed, either orally or in writing, before the parliament. Parliamentary immunity is linked to the duration of the mandate, and the decision to revoke the immunity of a parliamentarian rests within the parliament's jurisdiction.

Evaluation criteria on the status of parliamentarians also cover the monetary situation of parliamentarians. In particular, the criteria state that parliament should provide its members with appropriate monetary remuneration, a few material benefits to facilitate the fulfillment of their mandate, and reimbursement of expenses incurred during their mandate. Any form of compensation that parliament pays to its members must be allocated in a transparent manner on the basis of the parliamentarians' duties. In addition, to address conflicts of interest and corruption that are not already defined by the constitution or law, parliament can establish rules for its members on transparency and types of public and parliamentary activities. For instance, a procedure may be established for parliamentarians to declare their real estate assets. Also, parliamentarians should avoid situations in which their personal interests may affect the fulfillment of their duties. Applicable legislation must, therefore, prevent and penalize fraudulent practices of parliamentarians.

Finally, a criterion stipulates that a legal mechanism must be in place to protect the relationship between parliamentarians and interest groups. This mechanism can take the form of a public register of these interest groups and their activities.

Prerogatives of Parliament

The second component of the evaluation criteria relates to the prerogatives of parliament, which include the working methods of parliament, legislative functions, parliamentary monitoring, parliamentary committees, and international relations.

Working Methods of Parliament

The evaluation criteria reference document states that parliament—or each chamber, in some cases—must prepare, adopt, and amend its regulations in accordance with the constitution. Moreover, parliament must take meaningful measures to establish and maintain a balance of gender in its different branches at all levels of responsibility.

To ensure its proper operation, parliament must designate a chair and at least one vice chair according to the terms of its regulations. Parliamentary meetings must be held at sufficiently regular intervals to promote the efficiency and accountability of parliament. Parliament must also establish procedural rules to facilitate the holding of ordinary or extraordinary sessions, as well as the conditions under which the executive authority or fewer than all parliamentarians can meet.

Parliament must be able to participate in setting its agenda and allocating the time assigned to each of the agenda items for review. The agenda must be entrusted to a parliamentary branch, and parliament must inform its members well in advance about the timing and agenda of its meetings. In addition, a time-table for the legislative work must be established to strengthen predictability. The agenda must ensure that the bills and draft legislation are reviewed within a reasonable allotted time and that they are fully discussed. Finally, parliamentarians or members of the elected house of parliament must be able to submit proposals for legislation as well as amendments.

Legislative Functions

With regard to the legislative functions of parliament, the evaluation criteria provide that parliament must pass all legislation and the budget and that any exception to this rule must be clearly stated. Parliament must be able to adopt resolutions without notice and take a stance on a few topics of general interest. Moreover, parliament must have a clearly established legislative procedure, which governs the filing of proposed laws, their review, and their enactment.

In a bicameral parliament, each chamber's role must be clearly defined, and a conciliation procedure must exist in the case of an absence of agreement between the two chambers. In addition, an independent judicial branch should be responsible for ensuring, through constitutional scrutiny, that legislation is in accordance with the constitution.

With regard to parliamentarians, the evaluation criteria state that they must be able to table amendments, subject to the application of the rules governing their admissibility. Indeed, specific regulatory provisions should regulate the order of the amendments and the terms of their discussion to facilitate a clear organization of the debates and to encourage the expression of all views. Parliament must therefore establish and follow clear procedures that structure the course of parliamentarians' debates and must determine the order of precedence of the motions filed by members to provide opportunities for discussing bills and proposed laws prior to their vote. Only members can vote in parliament and, except as clearly stated, the votes in plenary sessions must be made public. The vote must be of a personal nature and, except for derogation clearly provided by law, delegating the right to vote is prohibited.

The criteria also stipulate that constituents must be involved in the legislative process through their parliamentary representatives. Constituents must be informed of the issues under discussion in parliament in a timely manner. In short, debates on bills and proposed laws must be open to the public, and all information regarding any legislation must be provided to all parliamentarians and all citizens.

Parliamentary Monitoring

A third aspect of parliamentary prerogatives reviewed in the evaluation criteria is parliamentary monitoring procedures. Parliament must be able to control the government's actions and access valuable information so that it can effectively exercise its monitoring operations. To that effect, a rigorous and systematic procedure for posing questions, either written or oral, to the executive branch must be clearly defined. In addition to the supervision of ministries, the parliamentary monitoring function must include the supervision of public enterprises and government agencies, including those in the defense and national security sectors.

The evaluation criteria also stipulate that parliament should have sufficient time to examine and discuss the national budget. Parliamentary committees should enable all parliamentary groups, under the rules of parliament, to scrutinize government spending. Parliamentary committees, especially those in charge of reviewing government spending, must have access to all necessary documents, including testimonies of all the ministries' and government agencies' senior officials, to effectively monitor government spending.

Moreover, a nonpartisan and independent organization (for example, the Audit Office or Office of the Auditor General) must be in place and have the necessary authority and adequate resources to carry out monitoring and auditing functions. Parliament must seek support from this entity and receive reports within a reasonable time to ensure effective follow-up. Finally, all institutions must provide clear mechanisms to establish a balance between the legislative and executive powers.

Parliamentary Committees

A fourth aspect of parliamentary prerogatives covered in the evaluation criteria concerns parliamentary committees. The criteria specify that parliamentary regulation must provide for the possibility of establishing permanent or temporary committees. Under parliament's regulations, committees' meetings should be held in public, and the conditions in which committees can express themselves in public meetings must be specified. Any exception to this rule must be stipulated and clarified in the regulations. Moreover, the working and voting procedures must comply with parliamentary regulations, which, in turn, must accurately define the jurisdiction and composition of committees.

The criteria also stipulate that committee qualifications should be clearly defined to avoid a conflict of jurisdiction. The committee's composition must reflect as closely as possible that of parliament and must take gender into consideration. Under the mechanism defined in parliament's regulations, a committee should select or elect one chair and at least one vice chair. Committees must be able to use the services of experts, and anyone called for a hearing by a commission of inquiry must be given a form of protection.

Finally, the criteria specify that parliament must submit draft bills and legislation to a committee for review and that any exception to this rule must be provided in parliament's regulations. Thus, the committees are called on to review and have the power to amend draft bills and legislation submitted to them. Only the committee's members can participate in internal votes. Last, committees can hold hearings and request any document they deem necessary for the efficient operation of their work.

International Relations

A final aspect of parliamentary prerogatives covered in the evaluation criteria is international relations. The criteria stipulate that, in parliamentary diplomacy, the delegations must reflect as closely as possible the composition of parliament and, in particular, take gender into account. Moreover, parliamentarians can participate in organizations or events allowing them to exchange their experiences with colleagues from other parliaments. They must be able to participate in missions with other parliaments and host foreign parliamentary delegations. Parliament must meet its obligations incurred with the international parliamentary institutions.

Parliaments can participate in international affairs and strengthen the parliamentary component of regional and international organizations. To this end, parliaments should have the information, organization, and resources required to undertake research on international issues. Moreover, parliamentarians should be included in government delegations during international missions or negotiations. With regard to assistance and cooperation, parliaments must be able to provide technical assistance to other parliaments to the extent of their ability. Similarly, parliamentary members and staff members must have the right to receive technical assistance.

Organization of Parliament

A third component of the evaluation criteria for democratic parliamentary procedures concerns parliamentary organization. This section reviews the four main aspects of parliamentary organization: status of political parties, parliamentary groups, and the opposition; status of the administrative staff; budget; and material means.

Status of Political Parties, Parliamentary Groups, and the Opposition

The first aspect of parliament's organization discussed in the evaluation criteria is the status of political parties, parliamentary groups, and the opposition. In particular, the criteria reference document notes that private and public funding of political parties, whenever it exists, must be based on transparent criteria. Equitable access to public funding must be provided, and a competent and independent judicial authority must ensure the monitoring of all funding.

With regard to parliamentary groups, they must obtain a legal status or another form of recognition. The criteria for formation of a parliamentary group, as well as its rights and responsibilities in parliament, must be clearly established in parliament's regulations. All parliamentary groups have the right to register their comments on the agenda, express their opinion, and propose amendments to bills. Parliament must therefore provide appropriate resources and adequate infrastructure to parliamentary groups.

Status of the Administrative Staff

A second aspect of parliament's organization that is governed by the evaluation criteria concerns the administrative staff. Parliament must rely on a permanent, professional, and nonpartisan administrative staff to provide impartial assistance to parliamentarians in a number of areas. Parliament must be independent of the executive branch, have control of the parliamentary services, and determine the conditions of staff recruitment and employment.

The administrative staff members must demonstrate impartiality and professional discretion to fulfill their role. In addition, they must be clearly distinguished from political staff members (people hired by and at the exclusive service of a parliamentary or a political group). Furthermore, the representation of women must be ensured at all levels of the hierarchy of the parliamentary administration.

With regard to the recruitment and promotion process, the criteria stipulate that parliament must have the resources to recruit a parliamentary administrative staff corresponding to its needs. The salary scale for staff members must be in accordance with that of government employees. Moreover, the recruitment and promotion of staff members must be conducted according to a fair and transparent selection process.

Finally, staff members must enjoy statutory protection from any form of undue political pressure. A mechanism should therefore exist to prevent, detect, and prosecute administrative staff or political staff of parliament who are engaged in fraud or corruption.

Budget
A third aspect of parliament's organization reviewed in the criteria is the budget. Parliament alone can determine and vote on its own budget, and the executive branch should not be the judge of appropriating the required means to parliament to fulfill its role.

Material Means
A final aspect of parliament's organization is material means. In particular, the evaluation criteria stipulate that parliament must receive adequate physical and material infrastructure so that its members can fulfill their mandate in satisfactory conditions.

Parliament and Communication
This section reviews the fourth and final component of the evaluation criteria, parliament and communication. The criteria reference document focuses on two key aspects of this component: access to parliament and dissemination of parliamentary information.

Access to Parliament
The first aspect of parliamentary communication raised by the evaluation criteria is access to parliament. Specifically, parliament must ensure that the media have rights to access all parliamentary public activities without interfering in parliament's proper functioning. The media's access to parliament must be made on a nonpartisan and transparent basis.

Moreover, parliament must be accessible to the public, provided that such accessibility does not undermine public safety and the parliamentarians' work obligations. Plenary meetings of parliament must be public, and parliament must have the means to facilitate citizens' understanding of its work. Finally, if the constitution or parliamentary regulations require the use of multiple working languages, parliament must make reasonable efforts to ensure mutual understanding among parliamentarians.

Dissemination of Parliamentary Information

The second aspect of communication concerns the dissemination of parliamentary information. Parliament must help develop a spirit of tolerance and promote democratic culture through education and training to raise the awareness of public officials, political actors, and citizens about the ethical requirements of democracy and human rights.

Laws, bills, and proposed legislation; committee reports; and any other parliamentary document provided for in parliamentary regulations must be made available to the public. Finally, parliamentary institutions must be more transparent and encourage the dissemination of their work through the use of communication tools, thereby fostering citizens' access to parliamentary information.

Challenges of Developing Evaluation Criteria for the Francophone World

Developing a comprehensive range of evaluation criteria has led to reflection on what priorities to establish. According to Louis Massicotte, a professor of political science at Laval University and the first holder of the Research Chair on Democracy and Parliamentary Institutions, this approach has numerous pitfalls. First is the risk of ethnocentrism, which is to take one's own national system as a point of reference. There is also the risk of overstating the cultural variable and having a conception of democracy dictated by the French model in opposition to others. A final potential pitfall is perfectionism: stakeholders may design the contours of democracy of the future, but for many countries, achieving democracy in the second half of the 20th century is still difficult.[3]

Establishing democratic criteria that can garner the majority of members' support has required a number of debates and amendments to take into account, as much as possible, country-specific cultural heritage. Thus, although consensus was reached on a number of universally recognized criteria, others gave rise to more debate. The rest of this section focuses on five main issues that emerged during the development process.

First, a fine balance was revealed between parliamentarians' right to freedom of expression and party discipline. This issue was particularly pertinent to the difficult debate on "crossing the floor," which occurs when a parliamentarian elected under one party banner voluntarily quits his or her party for another while a session is in progress. Generally, parliaments of the north run against those of the south over their perception of this change of allegiance.

For some parliaments, a parliamentarian's defection can cause problems for both the party and the parliamentarian's constituents. In fact, it can even change the outcome of an election. Therefore, a number of parliaments have adopted antidefection measures whereby any parliamentarian who crosses the floor loses his or her seat. However, some countries find these measures unacceptable because they compromise the parliamentarian's independence. These different approaches can be explained by diverse realities across countries: in some countries, this practice is limited primarily to individuals; in others, the scope is entirely different. In developing the evaluation criteria, it was not possible to find a solution to this matter that satisfied everyone. In recognition that more reflection is still needed, the Parliamentary Affairs Committee has taken over this theme, and a joint report on crossing the floor has since been produced by the National Assemblies of Quebec and Burkina Faso.

A second issue that caused debate was whether stakeholders could talk about a free, fair, and transparent election if discrimination exists with regard to gender, race, religion, language, sexual orientation, economic situation, or disability toward those standing for election. Although some of these criteria are now universally recognized, others still clash with cultural considerations, such as discrimination based on sexual orientation.

The issue of transparency in political life raised a third debate. With a view to greater transparency, many countries made it mandatory for parliamentarians to declare their financial and real estate assets. A number of parliaments have expressed reservations about this requirement, which has led the parliamentarians of APF branches to question whether such a procedure should be considered an essential criterion for good governance or a matter for each parliament to decide.

Although it was unanimously agreed that parliaments must have a qualified staff independent of the executive branch to effectively fulfill their role, a fourth conflict transpired around what universal criteria are needed to guarantee a democratic recruitment process. Consensus is still lacking about whether the recruitment and promotion of parliamentary staff members should be done through a merit-based competition and whether it should have fair and transparent criteria.

A final debate emerged with regard to the problems that multilingual parliaments face. However, the issue was addressed in discussions about whether parliamentary institutions should be required to use multiple working languages to ensure understanding among their members and what criteria should be used (equity, equality, or proportion).

Conclusion

The research presented in this chapter is a collection of progressive criteria and objectives for APF parliamentary members to strive for. Above all, it is a constructive document that aims to form the basis of a regular, targeted, and progressive process. Although the path to access this democratic ideal may vary according to each country's cultural, historical, economic, and social specifics, it will find its meaning in the desire of the francophone world to constantly improve the functioning of its parliaments.

Developing the evaluation criteria has enabled APF parliamentarians to share their experiences and look objectively at the best ways of serving democracy. The debates have led parliamentarians to better understand the meaning of shared commitment to democracy while respecting their particular paths. More generally, in identifying the criteria against which parliamentary democracy can be gauged, the APF was able to mobilize and reflect on the best means of ensuring that parliament, as an institution, works and has legitimacy.

The APF has endeavored to promote the evaluation criteria by sharing the reference document with all parliamentary branches and inviting them to use it in their strategic planning or for parliamentary reforms. The document is also available on the APF's website (http://apf.francophonie.org/) and the website of Agora (http://www.agora-parl.org/), an organization that aims to pool all available resources on parliamentary development. The APF has also committed to reassess these democratic criteria periodically.

Notes

1. This chapter is based on and updates APF (2009b), which was adopted in Paris at the 35th Session of the APF in July 2009.

2. Excerpts from OIF reference documents were also taken into account. Examples include the Bamako Declaration, which addresses democratic practices, rights, and freedoms in the francophone world (adopted November 3, 2000, Bamako), and the Declaration of Saint Boniface, which concerns conflict prevention and human security (issued by the Ministerial Conference of La Francophonie on May 13–14, 2006, Saint Boniface, Canada).

3. Comments are from an interview with Louis Massicotte (APF 2009a).

References

APF (Assemblée Parlementaire de la Francophonie). 2009a. "La réalité démocratique des parlements." *Parlements et Francophonie* 17: 8–9.

———. 2009b. "La réalité démocratique des Parlements: Quels critères d'évaluation?" Text adopted by the 35th session of the APF, Paris, July. http://apf.francophonie.org /IMG/pdf/La_realite_democratique_des_Parlements_-_Quels_criteres_devaluation _-_Geneve.pdf.

CHAPTER 5

The Parliamentary Confederation of the Americas: Toward the Consolidation of Parliamentary Democracy

Jacques Chagnon*

Introduction to COPA

The Parliamentary Confederation of the Americas (COPA) is an organization that brings together the congresses and parliamentary assemblies of unitary, federal, federated, and associated states; the regional parliaments; and the interparliamentary organizations of the Americas.

COPA was founded in response to the first Summit of Heads of State and Government of the Americas, held in Miami in 1994. The National Assembly of Quebec had the idea of bringing together some 300 federal parliamentary assemblies of the continent's national states, federal and regional parliaments, and interparliamentary organizations.

COPA was officially launched in 1997, when more than 400 parliamentarians from 28 countries of the Americas met in Quebec. At that time, they agreed on the need to create an autonomous representative interparliamentary forum for the continent that would be pluralistic and nonpartisan. During the meeting of the new organization, parliamentarians would address the political, social, environmental, and cultural effects of the integration process in the Americas. Three years later, COPA had a permanent structure, statutes, and rules of procedure.

COPA's mission is defined in its statutes:[1]

COPA encourages networking between the parliamentary assemblies as a means of enriching interparliamentary dialogue, and fosters the adoption of measures to ensure that our continent remains a peaceful zone founded on the principles of representative and participative democracy and social justice, the protection of

---

*Jacques Chagnon is the President, National Assembly of Quebec and the former President, COPA.

individual rights, gender equity and the forms of economic integration or complementarity that best suit our respective countries or groups of countries.

The organization has six permanent thematic working committees, one of which is the Committee on Democracy and Peace,[2] whose mandate is to examine issues including the strengthening and promoting of democracy and the rule of law. Over the past 10 years, the committee has implemented a program of election observation missions that contributes to the achievement of COPA's goals in promoting an active contribution by parliamentarians of the Americas to the development and strengthening of democracy in the Americas.

Development of the Benchmarks

If the legal and transparent electoral process is to be a cornerstone of a healthy democracy, it must also be reflected in the work of parliamentarians and the institutional framework within which they exercise their functions. Indeed, the responsibilities of parliamentarians as representatives of the citizens, legislators, or overseers of the executive power can be adequately met only in a context where certain standards are observed and where consensual constitutional and statutory guarantees exist. Hence, the committee is engaged in a process to develop benchmarking criteria for the parliaments of the Americas. This work began at the International Conference on Benchmarking and Self-Assessment for Democratic Parliaments, held in Paris, March 2–4, 2010.

The secretariat of the committee undertook an extensive reflection on the question of benchmarking for democracy and the specifics of the Americas. Indeed, because of its inclusiveness, COPA has a wide range of parliamentary systems, including parliaments and legislatures of federated states. It was necessary to find benchmarks on which members from British parliamentary systems and from presidential systems could agree. The continent's immense diversity must be both recognized and celebrated. Thus, the goal was certainly not to "standardize parliamentary systems or promote one single model of 'best practices'" (COPA Committee on Democracy and Peace 2011, 5), but to provide parliaments with a reference tool on parliamentary democracy.

As a premise to that reflection, COPA has assumed the existence of a consensus regarding the fact that representative democracy constitutes the most appropriate system of government, with the legislature being essential to the existence of a dynamic democracy.

During the development of COPA's benchmarks, the work already done by the Assemblée Parlementaire de la Francophonie (Parliamentary Assembly of La Francophonie), the Commonwealth Parliamentary Association, the Inter-Parliamentary Union, and the National Democratic Institute for International Affairs was recognized. Building on those efforts, but recognizing a distinct Americas perspective, COPA developed its own criteria, comprising four main sections: elections and the status of parliamentarians, parliamentary prerogatives, organization of parliament, and parliamentary communications (see annex 5A).

The first section focuses on the rules governing the electoral process, the eligibility of candidates, and the status of parliamentarians. These standards are designed to ensure that an elected parliament is representative from the sociological and political points of view. They also aim to identify the rights and limits of parliamentarians and to promote equal opportunities.

The second section defines all the areas that should be the responsibility of a parliament. It focuses on the organization of parliamentary work, parliament's legislative functions, parliament's function as public protector, parliamentary oversight, committees, and international relations.

The third section touches on everything that relates to the organization of parliament. According to standards and consensual democratic values, certain procedures and resources to guarantee the independence and autonomy of parliament, such as control of its budget, should be given special attention to ensure the effectiveness of parliamentary work. A neutral and qualified administrative staff, separate from government public service, and an adequate infrastructure are other ways to ensure the democratic character of an institution.

The fourth section, dealing with the parliamentary communication, lists the means available to ensure compliance with the three core values. First, a parliament must be transparent in the conduct of its activities. Second, it must ensure public participation in its work. Third, parliamentarians must be honest and accountable for their actions toward the electorate. Promoting freedom of expression, interaction with civil society, and dissemination of parliamentary information contributes to the achievement of these values.

The Adoption and Looking Forward

After a workshop in Quebec City to review and discuss the COPA criteria, some revisions were made, and the General Assembly of COPA adopted the criteria on September 9, 2011.

COPA is the only organization in the Americas that can establish its work on benchmarks of parliamentary democracy with a broad consensus across North and South America. With these criteria, the parliaments, congresses, and assemblies in the Americas now have a tool that will enable them to undertake a self-assessment and reform work that will help increase their legislative and democratic efficiency and thus better meet the citizens' ever-increasing and legitimate expectations.

The Committee on Democracy and Peace is supporting local initiatives related to the democratic strengthening of parliaments. A self-assessment document complying with COPA's benchmarks has been produced to help parliamentarians think about the functioning of their parliament. COPA's benchmarks can therefore serve as a tool and basis for analysis in this area.

With this major project, COPA intends to take an active part in the global discussion on strengthening parliaments and legislatures. With the adoption of these criteria, COPA welcomes the opportunity to showcase the specific features of parliaments in the Americas and is in the process of conducting a wide distribution of its benchmarks for the benefit of its members and of the global parliamentary community.

Annex 5A: Recommended Benchmarks for the Parliaments of the Americas

Parity between men and women is a fundamental benchmark of democracy.

1. Elections and the Status of Parliamentarians
   1.1 Elections
      1.1.1 The Constitution of the State must include basic rules to govern elections.
      1.1.2 Parliamentarians must be elected through universal suffrage, by free, direct, and secret ballot. However, in the case of a bicameral parliament, the second chamber may be governed by special rules provided for in the constitution or the legislation of the country concerned.
      1.1.3 Legislative elections must meet international standards for free, genuine, and transparent elections.
      1.1.4 The integrity and independence of the body that manages and supervises elections must be guaranteed with respect to its composition, mandate, powers, and budget.
      1.1.5 Discussion, research, and consultation must be encouraged to achieve an electoral system and electoral structures that enjoy broad support within society.
      1.1.6 To foster accountability, elections must be held at regular intervals. A legislature must be of limited duration and be followed by new elections.
      1.1.7 In order to foster better representation of social diversity, the participation of persons from underrepresented groups (e.g., young people, members of minorities, immigrants, and persons with disabilities) must be encouraged.
      1.1.8 The principles of fair competition and equality must be observed, and general standards of conduct for political actors must be defined during election campaigns.
      1.1.9 States must adopt legislation to govern the financing of political parties and election campaigns, and establish an independent body to ensure compliance with such legislation. Each party must develop internal by-laws to ensure compliance with legislation respecting the fair and transparent financing of election campaigns.
      1.1.10 Regional and global networks for sharing expertise and developing standards must be promoted.
      1.1.11 Legislation must allow international observers to conduct free and independent missions.

1.2 Eligibility and Representativeness

1.2.1   Restrictions on candidate eligibility must not be based on gender, religion, economic status, race, physical disability, or private life considerations.

1.2.2   Notwithstanding the preceding clause, special measures may be taken to ensure the representation of national or regional diversity and its components.

1.2.3   Electoral processes must be fair and guarantee that no voter, candidate, or party is penalized or discriminated against.

1.2.4   Seats must be divided among the parties in a manner that reflects as faithfully as possible the votes obtained by each party.

1.3 Status of Parliamentarians

1.3.1   Incompatibility

1.3.1.1   Incompatible parliamentary offices must be defined by law.[3]

1.3.1.2   In bicameral parliaments, parliamentarians may not be members of both chambers simultaneously.

1.3.1.3   A specific procedure must be established to monitor and sanction incompatibilities.

1.3.2   Parliamentary Immunity and Privilege

1.3.2.1   Parliamentarians must enjoy immunity for words spoken in the performance of their duties. Parliamentarians cannot be prosecuted, sued, wanted by the authorities, arrested, mistreated, detained, judged, or imprisoned after expressing opinions verbally or in writing before parliament or after voting in the performance of their duties.

1.3.2.2   Parliamentary immunity may not be used to place parliamentarians above the law.

1.3.2.3   Parliamentary immunity does not extend beyond a parliamentarian's term of office. However, former parliamentarians continue to enjoy protection for their term of office.

1.3.2.4   Parliament has exclusive jurisdiction to lift the immunity of a parliamentarian.

1.3.2.5   Parliamentarians must be able to perform the duties of office in accordance with the constitution, free from any undue influence or pressure.

1.4 Individual Rights of Parliamentarians and Party Discipline

1.4.1   Parliamentarians may only be expelled from their party in accordance with the party's internal by-laws, which must guarantee fair treatment, including the right to defend oneself.

1.4.2   Expulsion from a party must not automatically result in the loss of a parliamentarian's seat, or a reduction of his or her term, in violation of the right to free expression.

1.4.3   Only parliament may decide to exclude a parliamentarian from parliament under established rules, which must guarantee fair treatment, including the right to defend oneself.

1.4.4   The right of freedom of association exists for parliamentarians, as for all people.

1.5 Material Resources Provided to Parliamentarians

1.5.1 Indemnities

1.5.1.1   Parliament must provide parliamentarians with appropriate and fair remuneration, proper material infrastructure, and reimbursement for expenses incurred in the performance of their duties.

1.5.1.2   Any form of compensation paid to parliamentarians by parliament must be allocated in a transparent manner on the basis of the duties performed.

1.5.2 Conflict of Interest and Corruption

1.5.2.1   Parliament must establish rules, applicable to all parliamentarians, to govern transparency and the conduct of public and parliamentary activities.

1.5.2.2   There should be a legal mechanism to govern relations between public office holders and interest groups. The mechanism may be a public register of such interest groups and their activities.

1.5.2.3   Conflict of interest rules must be established to foster the independence of parliamentarians as regards private interests and undue political pressure.

1.5.2.4   Parliamentarians must avoid placing themselves in situations in which their personal interests may influence the performance of their duties.

1.5.2.5   A financial asset and business interest declaration procedure must be established for parliamentarians.

1.5.2.6   There must be legislation to prevent and sanction fraudulent practices by parliamentarians.

1.5.2.7   Preventive and repressive measures to fight corruption must be reinforced and enforced. Independent disciplinary bodies must be put in place to investigate corruption.

1.6 Resignation

1.6.1   Parliamentarians must be able to resign their seat at any time.

1.6.2   A replacement procedure must be established to fill vacant seats.

2. Parliamentary Prerogatives

2.1 Organization of Parliamentary Proceedings

2.1.1 General

2.1.1.1   Only parliament—or, as the case may be, each of the houses of parliament—may adopt or amend its rules of procedure.

2.1.1.2 The rules of procedure of parliament—or, as the case may be, of each of the houses of parliament—must be consistent with the constitution.

2.1.1.3 Parliament must take special measures in order to establish and maintain an equal proportion of women and men at all levels of responsibility throughout its organization.

2.1.2 Presiding Officers

2.1.2.1 Parliament—or, as the case may be, each of the houses of parliament—must elect or select a presiding officer and at least one deputy presiding officer pursuant to criteria and procedures clearly defined in its rules of procedure.

2.1.3 Legislative Sessions

2.1.3.1 Parliament must meet regularly, at intervals sufficient for it to fulfill its responsibilities.

2.1.3.2 Parliament must establish procedures for calling itself into regular or extraordinary session.

2.1.3.3 Provisions allowing the executive branch or a group of members to convene parliament must be clearly specified.

2.1.4 Plenary

2.1.4.1 The plenary must be organized in such a way as to allow enough time for the items on parliament's agenda to be examined.

2.1.4.2 Interference between the timing of the plenary and other parliamentary organs must be minimized.

2.1.5 Parliamentary Agenda and Calendar

2.1.5.1 Legislators must have the right to vote on the agenda and the time allowed for each item.

2.1.5.2 Parliament must give its members and the public sufficient advance notice of meetings and the agenda for the meetings.

2.1.5.3 A calendar of legislative work must be set so that the legislative schedule is known.

2.1.5.4 The agenda must ensure that proposed legislation is carefully examined in a reasonable timeframe by parliamentarians.

2.2 Legislative Functions

2.2.1 General

2.2.1.1 Members of parliament or of the elected house must have the right to introduce legislation and amendments.

2.2.1.2 All legislation, as well as the budget, must be passed by parliament. Exceptions to this rule must be clearly laid down.

      2.2.1.3 Parliament must have the power to adopt resolutions without advance notice and to take a stand on certain issues of general interest.

      2.2.1.4 Parliament must have the prerogative, under specific legal criteria, to delegate legislative functions to the executive branch for a limited period of time and with a view to achieving a clearly defined goal.

2.2.2 Legislative Procedures and Bicameralism

      2.2.2.1 Legislative work must be governed by a clear set of rules that cover the introduction of bills, their consideration, and their enactment.

      2.2.2.2 In a presidential system, parliament must have the right to override a veto of the executive branch.

      2.2.2.3 In a bicameral parliament, the role of each of the houses must be clearly defined.

      2.2.2.4 In a bicameral parliament, a conciliation process must be in place to resolve potential disagreements between the two houses.

2.2.3 Constitutionality of Legislation

      2.2.3.1 An independent judiciary must be made responsible for constitutional review, that is, for verifying whether laws that have been enacted are consistent with the constitution.

2.2.4 Power of Amendment

      2.2.4.1 Every parliamentarian must have the right to propose amendments, in accordance with the rules governing their admissibility.

      2.2.4.2 In order for debate to be organized and all opinions expressed, the order of amendments and the terms for discussion of amendments must be governed by strict regulatory provisions.

2.2.5 Debates

      2.2.5.1 Parliament must establish and follow clear procedures for structuring debate and determining the order of precedence of motions introduced by members.

      2.2.5.2 Parliament must provide adequate opportunity for members to debate proposed legislation prior to a vote.

2.2.6 Votes

      2.2.6.1 Only members of parliament may vote in parliament.

      2.2.6.2 Except for certain clear exceptions, plenary votes must be public.

2.2.7 The Legislative Process and the General Public

      2.2.7.1 Citizens must be involved in the legislative process, through their representatives in parliament or alternative means.

2.2.7.2 The public must be made aware in a timely manner of the issues being debated in parliament. Enough information must be made available to allow civil society to express its opinions regarding bills.

2.2.7.3 Information regarding legislation must be accessible not only to all parliamentarians, but also to the general public.

2.2.7.4 Debates on proposed legislation must be open to the public at some stage in the legislative process.

2.2.7.5 In the absence of a referendum, amendments to the constitution must be approved by the members of parliament.

2.3 Parliamentary Oversight

2.3.1 General

2.3.1.1 Parliament must be empowered to oversee the actions of the government.

2.3.1.2 The government must provide parliament with sufficient information for it to exercise its oversight function effectively.

2.3.1.3 A rigorous, systematic procedure must be established to govern questions (both written and oral) addressed to the executive branch by parliamentarians.

2.3.1.4 In addition to its oversight of government departments, parliament must oversee publicly owned enterprises and government agencies, including those in the defense and national security sectors.

2.3.1.5 In presidential systems, where ministers are not members of parliament, nominations for high-ranking positions within the executive branch must be subject to parliamentary approval following an in-depth examination of the nominee's fitness for the post.

2.3.2 Budget Review and Financial Control

2.3.2.1 Parliament must be given sufficient time to review and discuss the budget.

2.3.2.2 The law must guarantee the right of parliamentarians to create commissions of inquiry. Such commissions must have the power to compel persons outside of parliament, including executive branch officials, to appear and give evidence under oath. Persons testifying before a commission of inquiry must benefit from a form of protection.

2.3.2.3 Parliamentary committees specifically tasked with reviewing government expenditures must, in accordance with parliament's rules of procedure, allow all parliamentary groups an in-depth review of government spending. They must have access to all necessary documents and the power to hear high-ranking officials within government departments and agencies.

2.3.2.4 An independent, nonpartisan body (a tribunal of accounts or auditor general) must be put in place and provided with adequate resources and legal authority to carry out oversight and audit functions.

2.3.2.5 This body must report to parliament in a timely manner so that follow-ups may be conducted effectively.

2.3.2.6 Parliament must have the power to solicit the help of this body.

2.3.3 Relationship with the Executive Branch

2.3.3.1 In Westminster-style parliamentary systems, clear mechanisms must be put in place to ensure a measure of independence between the legislative and executive branches.

2.3.3.2 In presidential systems, an appropriate level of coordination must be established between the legislative and executive branches. To that end, the creation of special coordinating bodies or committees may prove essential.

2.4 Parliamentary Committees

2.4.1 General

2.4.1.1 The rules of parliamentary procedure must provide for the creation of standing or temporary committees.

2.4.1.2 Where stated in the rules of procedure, the sittings of a committee must be public. Exceptions must be clearly defined and provided for in the rules of procedure.

2.4.1.3 Committee proceedings and voting procedures must be consistent with the rules of procedure.

2.4.1.4 The rules of procedure must clearly describe the mandate and composition of committees.

2.4.1.5 To avoid conflicts of jurisdiction, committees must have clearly defined areas of competence.

2.4.1.6 The conditions under which a committee may vote in public must be outlined in the rules of procedure.

2.4.2 Selection of Committee Members

2.4.2.1 The membership of a committee must reflect that of parliament as closely as possible, with special consideration given to gender.

2.4.2.2 Committees must select or elect a chair and at least one vice chair according to the method described in the rules of procedure.

2.4.2.3 Committees must have the power to hire experts.

2.4.3 Terms of Reference

2.4.3.1 Proposed legislation must be referred to a committee for consideration. Exceptions to this rule must be transparent, clearly outlined in the rules of procedure, and extraordinary in nature.

2.4.3.2 Committees examine the bills referred to them and have the power to amend them.

2.4.3.3 Committees have the power to hold hearings and to summon any papers and records they require.

2.4.3.4 Only the members of a committee, or authorized substitutes, have the right to vote in committee.

2.4.4 Decision Making

2.4.4.1 Whenever possible, committees must strive for consensus in decision making.

2.5 Public Protector

2.5.1 Parliament must also exercise the function of public ombudsman by creating an independent body with the power to receive complaints from citizens who believe that they have been unfairly treated by the state or one of its bodies and to watch out for and correct inequities, injustices, abuses, and violations of rights committed by the state or one of its bodies.

2.5.2 This body must be completely independent from the government.

2.5.3 It must have broad investigative powers.

2.5.4 It must be provided with the necessary resources and be cost-free for complainants.

2.5.5 It must be easily geographically and electronically accessible.

2.5.6 It must report to parliament and be accountable to it.

2.6 Fostering Political Appeasement

2.6.1 Parliament must at all times serve the public interest and protect the welfare of citizens. It is responsible for fostering political appeasement by supporting democratic institutions and processes throughout the country.

2.6.2 Parliament must help settle political conflict in its country democratically, through dialogue and compromise.

2.7 International Relations

2.7.1 Parliamentary Diplomacy

2.7.1.1 Delegations operating within the framework of parliamentary diplomacy must reflect the membership of parliament as closely as possible, with special consideration given to gender.

2.7.1.2 Parliamentarians may take part in opportunities to share their experiences with members of other parliaments.

2.7.1.3 Parliamentarians must be prepared to take part in missions to other parliaments and to welcome delegations of foreign parliamentarians.

2.7.1.4 Parliament must fulfill its obligations towards international parliamentary institutions.

2.7.2 Participation in International Affairs

2.7.2.1 Parliament may participate in regional and international organizations, particularly in order to strengthen the parliamentary component of these organizations.

2.7.2.2 Parliament must have access to the necessary information, organization, and resources for examining international issues.

2.7.2.3 Parliamentarians must have the opportunity to be included in government delegations during missions or international negotiations.

2.7.3 Participation in the Regional Integration Process

2.7.3.1 Mechanisms must be put in place to facilitate cooperation between parliaments, in order to make coexistence with a regional parliament possible.

2.7.4 Cooperation and Support

2.7.4.1 Parliaments must be prepared to offer the best possible technical assistance to other parliaments.

2.7.4.2 Members of parliament and parliamentary personnel must have the right to benefit from technical assistance.

3. Organization of Parliament

3.1 Status of Political Parties[4]

3.1.1 General

3.1.1.1 Any conditions on the legality of political parties must be narrowly drawn in law and must be consistent with the International Covenant on Human and Political Rights.

3.1.1.2 Where it exists, public and private funding of political parties must conform to norms of transparency and accountability. A competent, independent judicial authority may oversee such funding. Equal access to public funding must be assured.

3.1.1.3 Parliament must encourage political parties to base their by-laws on principles of due process, clarity, transparency, and accountability.

3.1.2 Functions of Political Parties

3.1.2.1 Political parties may promote democratic values, human rights, tolerance, and the right to dissent.

3.1.3 Rights and Obligations of Political Parties

3.1.3.1 Political parties must be legally recognized and their legal existence certified by the state.

3.1.3.2 Political parties must be free to organize as they see fit, so long as they do not undermine the fundamental rights of members or other citizens, or run counter to the principles of the rule of law.

3.1.3.3 Political parties have a duty to act within institutional channels, using peaceful means to promote and achieve their political vision and objectives. Their actions vis-à-vis other parties must be respectful of democratic rules and procedures.

3.1.3.4 Political parties must uphold democracy within their organization, that is, they must adhere to democratic procedures and protect the fundamental rights of their members.

3.2 Status of Parliamentary Groups

3.2.1 Parliamentary groups must be granted legal status or some other form of recognition.

3.2.2 The criteria for forming a parliamentary group, as well as the rights and responsibilities of such groups, must be clearly stated in the rules of procedure.

3.2.3 All parliamentary groups have the right to place items on the agenda, to take part in debates, and to propose amendments to bills.

3.2.4 Parliamentary groups must be provided with adequate resources and facilities according to a clear, transparent, and equitable formula.

3.3 Status of the Opposition

3.3.1 The role of the opposition must be seen as beneficial to the democratic process.

3.3.2 Parliament must encourage conditions that guarantee a place for opposition parties in democratic parliamentary life.

3.4 Balancing Personal Life and Parliamentary Life

3.4.1 Parliament must be organized in such a way as to facilitate the participation of parliamentarians and allow them to fulfill their role while maintaining a balance between their parliamentary life and personal life.

3.5 Status of Administrative Personnel

3.5.1 General

3.5.1.1 The administrative management of parliament must be left to permanent, professional, nonpartisan personnel providing support for the various services.

3.5.1.2 Parliament must have control of parliamentary services and determine the terms of employment of its personnel independently from the executive branch.

3.5.1.3 Parliamentary personnel must carry out their functions with impartiality and mindful of their duty of restraint.

3.5.1.4 A clear distinction must be drawn and maintained between parliamentary service employees and political personnel (persons employed by a parliamentarian or parliamentary group and working exclusively for them).

3.5.1.5 Women must be adequately represented at all levels of parliamentary administration.

3.5.2 Recruiting and Promotion

3.5.2.1 Parliament must determine the terms for recruiting its permanent personnel independently from the executive branch.

3.5.2.2 Parliament must be provided with the resources necessary for recruiting the personnel it needs.

3.5.2.3 The recruitment and promotion of nonpartisan personnel must be based on merit, and the selection process must be fair and transparent.

3.5.2.4 When hiring or promoting employees, parliament must not discriminate on the basis of gender, religion, financial situation, race, or physical handicap.

3.5.3 Organization and Management

3.5.3.1 The status of parliamentary service employees must protect them from any form of undue political pressure.

3.5.3.2 Neither partisan nor nonpartisan personnel may have any legislative or procedural authority, including a vote in parliament.

3.5.3.3 Permanent and political personnel must be subject to a code of conduct. A mechanism must be put in place to deter, detect, and bring to justice any parliamentary employee engaged in fraudulent or corrupt practices.

3.6 Budget

3.6.1 Control of Parliament's Internal Budget

3.6.1.1 Only Parliament may determine and approve its budget, and the executive branch may not question the appropriateness of the means required by parliament for the exercise of its functions.

3.7 Material Resources

3.7.1 Facilities

3.7.1.1 Parliament must have access to the physical and material facilities necessary for its Members to carry out their functions under appropriate conditions.

4. Parliamentary Communications

4.1 Accessibility

4.1.1 The Media

4.1.1.1 Parliament must recognize access to information as a fundamental right of citizens. To allow this right to be fully exercised, parliament must ensure that the media are given appropriate access to the proceedings of parliament and its committees without, however, compromising its proper functioning.

4.1.1.2 Access by the media must be based on transparent, nonpartisan criteria.

4.1.1.3 Parliament must promote new information and communication technology and seek out ways in which technological advances could reinforce the democratic process and improve individual participation and decision making.

4.1.1.4 Parliament must promote freedom of expression.

4.1.2 The Public

4.1.2.1 The proceedings of parliament and its committees must be accessible to the public, as long as this accessibility does not interfere with public security or parliamentary business.

4.1.2.2 Plenary sessions of parliament must be open to the public.

4.1.2.3 Parliament must have access to resources for helping citizens understand its proceedings.

4.1.2.4 Parliament must ensure that the interaction between political parties and civil society is based on dialogue and cooperation.

4.1.3 Language

4.1.3.1 Parliament must facilitate the use of all working languages recognized by the constitution or in the rules of procedure, including simultaneous interpretation in debates and proceedings and the enactment of laws in all working languages.

4.2 Dissemination of Parliamentary Information

4.2.1 General

4.2.1.1 Key decision-making processes must be presented in detail when they are officially recorded.

4.2.1.2 Parliamentarians must disclose their assets before, during, and at the end of their term.[5]

4.2.2 Democratic Values

4.2.2.1 Parliament must foster a spirit of tolerance and promote all aspects of democratic culture in order to educate and raise awareness among public officials, political actors, and citizens about the ethical requirements of democracy and human rights.

4.2.2.2 Any restriction of freedom of expression must be prescribed by law. If restrictions prove necessary (for reasons of national security or to protect rights or reputations, for example), they must be proportional to their objectives.

4.2.3 Access to Legislation

4.2.3.1 Laws, proposed legislation, committee reports, and any other parliamentary document provided for by the rules of procedure must be made accessible to the public.

4.2.4 Access to Open Sittings and Committee Debates

      4.2.4.1 Parliament must encourage the use of widely available information and communication tools to broadcast its proceedings.

Notes

1. The quotation is from section 2 of the statutes. The full text of the statutes is available online at http://www.copa.qc.ca/eng/who/Statuts-COPA-a.pdf.

2. The others are the Committee on Economy, Trade, Labour, Competitiveness, and Trading Blocs; the Committee on Education, Culture, Science, and Technology; the Committee on Health and Social Protection; the Committee on the Environment and Sustainable Development; and the Committee on Human Rights, Aboriginal Peoples, and Citizen Security.

3. The term *incompatible* is used as defined by Merriam-Webster (2003, 630): "Incapable of being held by one person at one time—used of offices that make conflicting demands on the holder."

4. The term *political party* also refers to other political entities, such as citizen movements and associations.

5. The extent of public disclosure of assets depends on the standards adopted by each parliament.

References

COPA (Parliamentary Confederation of the Americas) Committee on Democracy and Peace. 2011. "The Contribution of Parliaments to Democracy: Benchmarks for the Parliaments of the Americas." Québec Secretariat of COPA, National Assembly of Québec. http://www.copa.qc.ca/eng/assembly/2011/documents/DOC-CDP-criteres -a-VF.pdf.

Merriam-Webster. 2003. *Merriam-Webster's Collegiate Dictionary.* 11th ed. Springfield, MA: Merriam-Webster.

CHAPTER 6

Benchmarking for Democratic Parliaments

Anthony Staddon and Dick Toornstra

Introduction

At the 2010 International Conference on Benchmarking and Self-Assessment for Democratic Parliaments, participants agreed that a democratic parliament "is one that is representative of the political will and social diversity of the population, and is effective in its legislative, oversight and representational functions, at the subnational, national, and international levels. Crucially, it is also transparent, accessible, and accountable to the citizens that it represents" (WBI and UNDP 2010, 3).

But how effective are parliaments in meeting such core values? Assessing parliamentary effectiveness requires some form of criteria and measurement of performance. In recent years, significant progress has been made in developing both benchmarks and self-assessment approaches. This chapter examines the rationale behind parliamentary benchmarks and self-assessment frameworks and makes some initial suggestions as to how they can be operationalized, depending on a parliament's development and resources.

This chapter is organized as follows: The next section discusses the merits of benchmarking for parliaments. The following section reviews existing benchmarking tools for legislatures. The subsequent section assesses how the use of benchmarks may vary across different types of legislatures. Finally, the last section concludes.

Merits of Benchmarking

Benchmarks for democratic parliaments are growing in popularity for several reasons. First, efforts are being renewed to build public confidence and strengthen the capacity of parliament to manage increasing demands and to assert greater

This chapter is a shorter version of a publication from the Office for Promotion of Parliamentary Democracy in the European Parliament titled *Benchmarking for Parliaments: Self-Assessment or Minimum Criteria?* (Staddon 2012).

institutional independence (Hubli 2009, 2010). At the same time, donors are now required to justify both their expenditure on parliamentary development and the effectiveness of these interventions. For interparliamentary organizations (IPOs), benchmarks provide an opportunity to codify programs and best-practice guides and to share experiences across parliaments (Hubli 2009, 2010). This exercise is particularly useful because members may be more open to receiving advice from their peers in IPOs.[1] Indeed, in the case of reform efforts made in Bermuda, members of parliament needed an independent platform on which to base changes and educate civil servants and the public about the basic needs of parliament (Smith 2010).

Most organizations involved in parliamentary benchmark exercises have similar overall objectives for their schemes of assessment. The Inter-Parliamentary Union (IPU) summarizes these objectives into two categories: (a) to evaluate parliament against international criteria for democratic parliaments and (b) to identify priorities and means for strengthening parliament. These basic objectives can then be further broken down into subobjectives or entry points for their use (table 6.1).

There are, of course, difficulties in applying benchmarks to legislatures. For instance, parliaments may be reluctant to measure their own work for fear of exposing bad practices or because of doubts about the practicality of the exercise. A review of developments in legislative oversight, for example, found that parliamentary committees seldom quantify information such as changes of legislation, cost savings, or improvements in service (CCAF-FCVI 2004, 10).

Moreover, the executive may view parliamentary benchmarking exercises in zero-sum terms, rather than the positive-sum goal of improved democratic performance. In this respect, a benchmarking exercise is likely to face problems similar to those faced by parliaments on a day-to-day basis. Fundamentally, political will and leadership must exist within parliament, often with support of the executive and outside agencies.

A country's historical and social context also has implications for the use of benchmarking exercises (box 6.1). Benchmarks need to be flexible to be relevant across the range of parliamentary and democratic models, but this flexibility inherently leads to complications. In particular, parliamentary stakeholders sometimes have contradictory understandings of what benchmarks are. For example, are they minimum standards, ideals, or goals? This ambiguity can lead to confusion as to how to position a parliament in relation to a given benchmark.

Systems of Benchmarking

Existing benchmarking tools for legislatures serve a range of objectives. Some sets of standards seek to codify good practices for self-assessments, whereas others seek to identify the minimum criteria for being a democratic parliament. The differences between frameworks mirror a larger debate on what constitutes democracy. For some, a democracy is a political system in which the principal positions of power are filled "through a competitive struggle for the

Table 6.1 Benchmarks and Standards: Summary of Possible Benefits

Parliaments and parliamentarians	International organizations and donor community	Civil society and general population	Academics
• To ensure the relevance and effectiveness of parliaments in the long term and to empower them to claim their proper place in the state's institutional order • To agree on overall results or objectives for legislative strengthening programs by engaging in a detailed level of analysis, introspection, and review • To help develop survey tools for members of parliament and staff members that measure attitudes, behaviors, and perceptions of the legislature's performance, thereby enhancing public confidence in parliamentary integrity • To enhance legislative transparency and accountability • To build political coalitions of interest: reform-minded legislators or staff members can use benchmarks to push for reform • To expose bad practices while keeping up to date with advances in parliamentary practice and procedures • To help prepare the parliamentary budget or strategic plan • To provide education and training, especially for new members of parliament, and to promote gender sensitivity in parliament • To enable parliamentary staff members to contribute their views more effectively and efficiently • To support requests for external assistance	• To use in designing parliamentary strengthening programs and in determining where to focus support • To enable interparliamentary organizations to codify their wider programs and best-practice guides and to share experiences of member parliaments • To design both qualitative and quantitative indicators that more accurately measure the effect of donor assistance on the performance of parliamentary institutions over time • To assist compliance with the principles of the Paris Declaration and Accra Agenda for Action • To ensure buy-in from legislatures for legislative strengthening programs • To ease the sensitivity sometimes evident in parliamentary assistance • To justify expenditure on parliamentary support programs	• To help promote change from outside the institution and to make a nongovernmental organization or civil society organization assessment of parliament • To manage increasing demands by building institutional capacity and helping influence the parliamentary budget or strategic plan • To use as an educational tool that (a) provokes wider debate about parliament and its role in consolidating democratic systems and (b) ensures greater public confidence in, and knowledge of, the legislature • To advocate for greater representation of women, minorities, and others • To promote gender sensitivity in parliament	• To increase academic interest in legislative development as a critical element of democratic institutionalization • To serve as a guide for evaluating the strengths and weaknesses of individual legislatures and to rank parliamentary power or effectiveness • To establish a set of democratic norms and values through which parliament operates • To encourage more comparative research on the use of different assessment frameworks

Source: These benefits have been compiled from the papers submitted and speeches delivered at the International Conference on Benchmarking and Self-Assessment for Democratic Parliaments, Paris, March 2–4, 2010.

Box 6.1 Country-Specific Factors That Can Affect Benchmark Exercises

Analysis of the country-specific context must include the following:

• Political background
• Constitutional and international rights and obligations
• Relationships between the parliament, the executive, and the judiciary
• Public perception and public access to parliament
• Socioeconomic, cultural, and traditional context

Source: IFES 2005, 7.

---

people's vote" (Schumpeter 1947, 269). For others, democracy has a broader definition: moving beyond free, fair, and competitive elections to encompass freedoms that make elections truly meaningful (such as freedom of organization) and institutions to ensure that government policies depend on the votes and preferences of citizens (Diamond 2002).

Many parliamentary organizations, such as the Commonwealth Parliamentary Association (CPA) use benchmarks as minimum standards rather than as questions.[2] Self-assessment tools are used to track a parliament's progress against an accepted standard or to support a request for external assistance. The self-assessment process is about fact seeking and may even be seen as prescriptive or normative because of the way the benchmarks are stated. The inherent risk of such a method is that benchmarks are too low; benchmarking is not useful if parliaments can easily meet the series of standards or if the benchmarks are simply an exercise in checking the boxes to gain international and domestic acceptance. One way to prevent these situations is to engage civil society and parliamentary monitoring organizations in discussions about benchmarks.[3] This approach may lead to a greater understanding of the constraints facing parliament and to broader support for parliamentary strengthening.

A second method for ensuring that benchmarks are useful is to go beyond the minimum requirements for a democratic parliament and actually codify good practices. The IPU adopts this approach.[4] A third approach is the assessment framework of the European Commission, which was developed for donors to engage with parliaments using parliamentary strengthening programs. The assessment framework is designed to identify focus areas for development through four steps: (a) pinpointing areas where a parliament is not currently performing aspects of its core functions, (b) understanding the possible underlying causes of these weaknesses, (c) identifying entry points of parliamentary development, and (d) designing context-specific parliamentary support programs.

A fourth method is a standards-based questionnaire developed by the National Democratic Institute for International Affairs (NDI) that compares individual legislatures to norms and basic functions of other parliaments and identifies best practices and lessons learned. The survey is designed to compare the perceptions of parliamentarians, parliamentary staff members, and representatives of civil society. NDI's questionnaire is unique in that it measures the perception gap between the real powers of the legislature and the powers that legislators exercise in practice—the gap between "having" and "using" power. It has been designed as a diagnostic tool to obtain "a clearer sense of the state of [the] legislature ... providing a foundation from which NDI, the legislature, and dedicated citizens can collaborate to create possible steps to further strengthen and enable the elected body" (NDI 2009, 1).

All methods assess the current state of a legislature against international criteria, thereby providing examples of issues to consider and stimulating debate about what kind of institution the organization should become. Benchmarking exercises are not designed to rank the legislature against others: the purpose is to improve the functioning of an individual legislature.

Moreover, the exercise must be repeated at regular intervals because developments and the context in which the parliament operates are dynamic. This ongoing examination is particularly important for methodologies that set minimum benchmarks, because the expectations of any democratic parliament should increase over a period of time.

Each of these different approaches emphasizes that the diagnosis of strengths and weaknesses and the establishment of development priorities is a process that belongs to parliament itself (though independent experts or consultants may be involved in carrying forward the process). Staff members should be involved in all exercises because they provide greater and sharper insights than parliamentarians in many jurisdictions.

The willingness of parliamentarians (and staff members) to engage in benchmarking or self-assessment exercises is seriously conditioned by the maturity of the legislature and the resources available to them. Experience to date suggests that even if willingness exists to work along these lines, some form of external encouragement or facilitation is also needed. The European Commission assessment framework is helpful in this regard because it details the different strategic entry points of intervention modalities for European Commission parliamentary support programs.

Parliamentary Entry Points for Benchmark and Assessment Frameworks

Because legislatures differ in terms of their institutional development and powers, variation in the use of the different benchmark and assessment frameworks is unavoidable. For example, a benchmark assessment in an advanced democracy is less likely to be externally driven or tied to a development program. Moreover, legislatures in small countries may operate as a mature parliament, but they may have greater difficulties than larger jurisdictions because of the small number of members available to participate. Furthermore, mature or advanced legislatures should aspire to the highest standards, and an approach that merely assesses whether a legislature meets minimum standards is likely to be less informative.

The different stages of parliamentary development can explain some differences in the benchmarks used. As with all such divisions, in these stages a degree of artificiality will exist, as well as some overlap. Yet for the purposes of this exercise, parliaments can be classified into the following three broad categories:

- *Emerging legislatures.* These parliaments are in the initial stages of setup or have been under way for a brief period.
- *Developing legislatures.* These parliaments have some experience with practice and procedure and more than the minimum level of competence in parliamentary responsibilities.
- *Mature legislatures.* These parliaments possess comprehensive technical, administrative, and political competences and meet at least some recognized international good practices.

Most parliaments view themselves as developing legislatures: not fully institu-tionalized but meeting some basic requirements of a democratic parliament and in the midst of a process of open-ended political change. Of course, legislatures may move backward as well as forward in their development. Just as there is no guarantee that any country moving away from dictatorship is in transition to democracy (Carothers 2002), there is no automatic linear progression in parlia-mentary development. A move backward may be induced by periods of political instability, conflict situations, or financial or economic pressures. Moreover, how institutionalized a parliament is will often depend on the size of the jurisdiction, its socioeconomic level, and its democratic maturity.

The choice of different types of benchmarks (from the broad IPU approach to the specific check-the-box CPA approach) can be puzzling if the distinctions among them are not clearly understood. One solution is that parliaments can assess themselves using both systems.[5] The IPU approach can be used to examine the legislature against the broader background of democracy in the country, and the CPA approach can then assist in standardization against international norms.[6]

Generally, however, the process of operationalizing any benchmarking scheme requires attention to four key questions:

- What considerations will affect the choice of benchmarking scheme and its operations?
- Should any benchmarks be prioritized over others?
- What level of implementation should be used? Is a minimum level of imple-mentation acceptable?
- What practical actions can be taken to meet each benchmark?

Mature parliaments may be most interested in best practices and innovative solu-tions to improve the quality, efficiency, and effectiveness of their core businesses. In contrast, emerging or developing legislatures are more likely to undertake a benchmark assessment for a specific purpose (for example, donor oriented), rather than as an exercise undertaken in the course of regular work. Emerging and developing legislatures will need to focus on areas where they have the best chance of getting results and move step by step under a practical plan of action to meet selected benchmarks. A sensible starting point for legislatures is to pri-oritize benchmarks that are common across the various approaches, because they will generally be accepted as having wider legitimacy. For that purpose, five broad themes have been identified across existing assessment frameworks: insti-tutional independence, procedural fairness, democratic legitimacy and represen-tation, parliamentary organization, and core legislative and oversight functions (box 6.2).

Although parliamentary benchmarks are based on a common minimum stan-dard, parliaments should be conscious of what is achievable and that an incre-mental, step-by-step approach is more likely to be successful and sustainable. CPA benchmarks relating to parliamentary committees serve as a prime example of the need to prioritize benchmarks: core benchmarks could involve the right to

Box 6.2 Assessment Frameworks for Democratic Parliaments: Areas of Consensus

Institutional Independence

Areas of consensus include parliamentary immunity, budgetary autonomy, control over staff, recourse to own expertise, sufficient resources to perform constitutional functions, adequate physical infrastructure, control over internal rules, and calling of extraordinary sessions.

Procedural Fairness

Areas of consensus include written procedural rules, plenary sittings in public, order of precedence of motions and points of order, meaningful opportunity for debate, use of official languages, right of all members to express their views freely, and arrangements to ensure that opposition and minority parties can contribute effectively to the work of parliament.

Democratic Legitimacy

Areas of consensus include democratic elections; election of the lower house through universal suffrage; regular periodic elections; and no restrictions on candidacy by race and gender, language, or religion.

Parliamentary Organization

Areas of consensus include right of legislatures to form committees; presumption that legislation is referred to committees; election of committee chairs and leadership according to procedures; right to form parliamentary party groups; right to a permanent, professional, nonpartisan staff; and protection of head of the nonpartisan service from undue political pressure.

Core Legislative and Oversight Functions

Areas of consensus include ability of the lower house to initiate legislation, rights to propose amendments and to amend legislation, right to consult experts and staff members on legislation, ability to hold public hearings or receive testimony from experts, right to subpoena or obtain documents, and methods for protecting witnesses.

Source: WBI and UNDP 2010.

form permanent and temporary committees; a balanced composition; and the power to summon papers, persons, and records. Benchmarks dealing with the legislature and the media could also be prioritized. Benchmarks stating that "the legislature shall be accessible and open to citizens and the media, subject only to demonstrable public safety and work requirements," and requiring the legislature to "ensure that the media are given appropriate access to the proceedings of the legislature without compromising the proper functioning of the legislature and its rules of procedure," apply to all legislatures (CPA 2006, benchmarks 9.1.1 and 9.1.2).[7]

Secondary benchmarks could include the right to consult or employ experts, which may prove difficult for capacity-constrained legislatures. The benchmark on the transparency of committee proceedings may also be a lower priority in some jurisdictions: although most international observers would agree that proceedings in parliament should be public,[8] this issue is still contested in some countries and regions.[9]

Many parliaments also agree on common benchmarks at the regional level. For instance, regional attempts have been made to codify benchmarks relating to gender equality.[10] However, the regional dimension also yields interesting differences and helps clarify which sets of benchmarks or standards relating to gender may be intended as a minimum standard[11] and which may be more aspirational in nature.[12] For example, the CPA benchmark stating that "restrictions on candidate eligibility shall not be based on religion, gender, ethnicity, race or disability," is accepted across all regions that have set their own benchmarks (CPA 2006, benchmark 1.2.1). However, the Southern African Development Community Parliamentary Forum has added creed and marital status to this list and has clarified that citizenship, age, or residency requirements are permitted (SADC PF 2010). Moreover, the Parliamentary Confederation of the Americas has added economic status and private life considerations to the CPA's original benchmark (COPA 2011; see also chapter 5 of this volume). Notwithstanding these regional differences, the original CPA benchmark can be seen as a necessary first step for all parliaments.

The preceding discussion has focused on the use of benchmarks. Both the IPU and European Commission approaches, which seek to codify good practices for purposes of self-assessment, are helpful for legislatures to identify their strengths and weaknesses (for instance, in the area of gender) and to formulate recommendations for reform. Mature legislatures may find this approach more useful or aspirational than meeting internationally agreed benchmarks.

Conclusion

How parliaments can improve their work and become more effective is now a focal point of international debate. A number of parliamentary assessment approaches exist,[13] and each approach is trying to do something different. Some provide minimum standards, whereas others draw on best practice and are more aspirational. However, the end goal—improving the performance of parliament and therefore the wider democratization process—is the same, and the methodologies used are often similar. Moreover, assessment frameworks typically evaluate parliament against international criteria but emphasize national ownership of the exercise. This provides a framework for parliamentarians to discuss the performance of their own legislature while engaging with other stakeholders.

Previous studies of benchmarks and assessment frameworks have revealed a broad consensus over many key areas of parliamentary practice, such as institutional independence and procedural fairness. These areas of consensus provide a convenient starting point for legislatures seeking to strengthen their performance. However, differences across frameworks have also allowed for wider debate and context to be studied. Interestingly, the CPA-derived minimum standards are now being applied within regions, which is encouraging the development of more aspirational benchmarks in some regions; however, it may also lead to the watering down of minimum standards agreed previously at the international level.

Last, we must recognize that different pressures and motivations will come into play, depending on the level of parliamentary development. Many legislatures will undertake a benchmark assessment for a specific purpose, perhaps donor oriented or because of a political desire to improve a parliament's functions and power, rather than as an exercise undertaken in the course of regular work. The IPU and European Commission approaches help legislatures ascertain where they are based in terms of their development. The IPU's Self-Assessment Toolkit for Parliaments (IPU 2008) is most useful for identifying strengths and weaknesses and formulating an action plan for development. In contrast, the CPA benchmarks help parliaments prioritize objectives and develop practical action plans according to their stage of development. Some legislatures may struggle to prioritize objectives because of the range of issues facing them or because of the difficulty in evaluating exactly where they stand. However, legislatures can address this obstacle by first undertaking the IPU exercise to identify the CPA benchmarks in which they are weakest. As such benchmarks become more widely accepted, the focus will be on more parliaments to start using these frameworks, and the role of international stakeholders should be to facilitate this process by sharing practical examples.

Notes

1. The continued development of IPOs is crucial to strengthening normative values and interparliamentary cooperation. The latter is becoming increasingly important as governments continue to establish a variety of formalized cooperation structures.

2. For more about the CPA, see chapter 3 of this volume.

3. See chapter 8 of this volume for more information about parliamentary monitoring organizations.

4. The IPU's approach is discussed in chapter 2 of this volume.

5. See chapter 10 of this volume, where members and staff of the Parliament of Sri Lanka have used both approaches.

6. Interestingly, both the use of the CPA benchmarks in Canada with a mixed group of parliamentarians and staff members and the use of the IPU framework in Sri Lanka with staff members only raised a point that is not specific in either set of benchmarks: the need for safeguards for the oversight of delegated or secondary legislation.

7. However, the benchmark stating that "the legislature shall have a nonpartisan media relations facility" may be a less immediate priority to emerging (and some developing) legislatures (CPA 2006, benchmark 9.1.3).

8. Issues of national security are usually excepted.

9. For example, holding committee hearings and votes in public is not common practice in Sri Lanka.

10. For instance, all assessment frameworks acknowledge that the legislature must not discriminate in the recruitment and promotion of staff. Indeed, the Southern African

Development Community Parliamentary Forum added a benchmark providing for equitable gender representation in the election of presiding officers (SADC PF 2010), whereas the Parliamentary Confederation of the Americas (COPA) includes benchmarks stating that parliament must maintain an equal proportion of women and men at all levels of its organization (COPA 2011). COPA and the Assemblée Parlementaire de la Francophonie (Parliamentary Assembly of La Francophonie) also add a requirement that delegations operating within the framework of parliamentary diplomacy must reflect the membership of parliament as closely as possible, with special consideration given to gender (APF 2009; COPA 2011).

11. Minimum thresholds may include candidate eligibility, the possibility of special measures, fair remuneration, adequate physical infrastructure, and no discrimination in the recruitment and promotion of staff members.

12. Aspirational benchmarks may include child care facilities, equitable gender representation in the election of presiding officers, special measures to establish and maintain an equal proportion of women and men at all levels of responsibility, and special considerations given to gender when selecting parliamentary delegations.

13. However, all are works in progress.

References

APF (Assemblée Parlementaire de la Francophonie). 2009. "La réalité démocratique des Parlements: Quels critères d'évaluation?" Text adopted by the 35th session of the APF, Paris, July. http://apf.francophonie.org/IMG/pdf/La_realite_democratique_des _Parlements_-_Quels_criteres_devaluation_-_Geneve.pdf.

Carothers, Thomas. 2002. "The End of the Transition Paradigm." *Journal of Democracy* 13 (1): 5–21.

CCAF-FCVI. 2004. "Parliamentary Oversight: Committees and Relationships." Background research paper for *Review of Recent Developments in Legislative Oversight in Britain and Australia, with Special Reference to Public Accounts Committees.* Ottawa: CCAF-FCVI.

COPA (Parliamentary Confederation of the Americas) Committee on Democracy and Peace. 2011. "The Contribution of Parliaments to Democracy: Benchmarks for the Parliaments of the Americas." Québec Secretariat of COPA, National Assembly of Québec. http://www.copa.qc.ca/eng/assembly/2011/documents/DOC-CDP-criteres -a-VF.pdf.

CPA (Commonwealth Parliamentary Association). 2006. "Recommended Benchmarks for Democratic Legislatures." CPA, London. http://wbi.worldbank.org/wbi/Data/wbi /wbicms/files/drupal-acquia/wbi/Recommended%20Benchmarks%20for%20 Democratic%20Legislatures.pdf.

Diamond, Larry Jay. 2002. "Thinking about Hybrid Regimes." *Journal of Democracy* 13 (2): 21–35.

Hubli, K. Scott. 2009. "Benchmarks and Standards for Democratic Parliaments: An Emerging International Consensus?" Presented at the Joint Inter-Parliamentary Union and Association of Secretaries General of Parliament Meeting, Geneva, October 22.

———. 2010. "Assessment Framework for Democratic Parliaments: Common Themes." Paper delivered at the International Conference on Benchmarking and Self-Assessment for Democratic Parliaments, Paris, March 2–4.

IFES (International Foundation for Electoral Systems). 2005. *Global Best Practices: A Model Annual State of the Parliament Report.* Washington, DC: IFES.

IPU (Inter-Parliamentary Union). 2008. "Evaluating Parliament: A Self-Assessment Toolkit for Parliaments." IPU, Geneva. http://www.ipu.org/pdf/publications/self-e.pdf.

NDI (National Democratic Institute of International Affairs). 2009. "NDI Standards-Based Questionnaire." NDI, Washington, DC.

SADC PF (Southern African Development Community Parliamentary Forum). 2010. "Benchmarks for Democratic Legislatures in Southern Africa." SADC PF, Windhoek. http://www.agora-parl.org/node/2777.

Schumpeter, Joseph. 1947. *Capitalism, Socialism, and Democracy.* 2nd ed. New York: Harper.

Smith, Jennifer. 2010. Speech at the International Conference on Benchmarking and Self-Assessment for Democratic Parliaments, Paris, March 2–4.

Staddon, Anthony. 2012. *Benchmarking for Parliaments: Self-Assessment or Minimum Criteria?* Brussels: European Parliament Office for Promotion of Parliamentary Democracy.

WBI (World Bank Institute) and UNDP (United Nations Development Programme). 2010. "Participants' Statement." International Conference on Benchmarking and Self-Assessment for Democratic Parliaments, Paris, March 2–4. http://www.ndi.org/files/Benchmarks_Conference_Participant_Statement_March2010.pdf.

CHAPTER 7

Parliamentary Benchmarks: A Requisite for Effective Official Development Assistance

Alice French

> *The budget is at the heart of efforts to improve governance and accountability in all countries. It is crucial in reducing poverty. The parliament can make an important contribution by expanding its oversight role throughout the budget cycle.*
>
> —Cheam Yeap, chairman of the Commission on Economy, Finance, Banking, and State Audit, Cambodia

Introduction

As the overseas aid dial nudges toward national budget support and effectiveness of aid is seen as being at least as important as the volume of aid pledged, international institutions such as the World Bank and bilateral donors of development assistance are looking to broaden the scope of aid-delivery programs in a way that enhances country ownership. Donors are increasingly allocating resources to develop the national institutions in aid-receiving countries as a means of both promoting mutual accountability to ensure aid efficiency and supporting long-term self-sustainability objectives. This approach replaces one that was mainly project driven and in which donors typically engaged only with the executive branch of government in recipient countries. More and more emphasis is being placed on donor support for the functioning recipient nations' domestic institutions, which are increasingly recognized as "vital allies for donor agencies in improving domestic accountability," parliament being foremost of these institutions (GOVNET 2011, 49). In this chapter, I examine the pivotal role that parliaments can play in ensuring that aid is used effectively and for the benefit of the general population. I also examine the need for the implementation of a set of

The author would like to acknowledge the input and guidance from Rasheed Draman, Mitchell O'Brien, Keith Schulz, and Rick Stapenhurst.

universally accepted standards or benchmarks by which democratic legislatures should abide. Such benchmarks will allow the functioning of parliaments to be strengthened and ensure that the concept of domestic accountability is fully realized through robust parliamentary oversight of the executive's deployment of aid income.

The first section of this chapter charts the evolving landscape of international aid assistance following a series of transformative Organisation for Economic Co-operation and Development (OECD) forums on aid effectiveness held in Paris, Accra, and Busan over the past decade. In particular, it looks at how the official development assistance (ODA) policies of major multilateral and bilateral providers have progressively changed toward general budget support in recipient countries coupled with the funding of parliamentary strengthening initiatives for improving governance and ensuring that unearmarked budget support is not wasted.

Second, the chapter explores the implications of the new thinking and modus operandi for multilateral donor organizations such as the World Bank, as well as for major bilateral international aid donors. Third, it details the essential role that parliaments in the aid-receiving countries play—or should be encouraged to play—in ensuring the effectiveness with which income from development assistance is used in the budget support model.

Finally, the chapter concludes with a discussion on the need for the international aid community—assisted by specialists in parliamentary best practice—to develop a set of universally recognized benchmarks. These benchmarks should cover not only the aspects of parliamentary functioning that are concerned with national budget oversight, but also other matters connected with the allocation, monitoring, and post implementation review of government expenditure funded from development assistance. It shows that implementation of such benchmarks is needed both to determine the capacity of national parliaments to oversee government activities involving aid income and to provide a framework for effecting parliamentary improvement strategies in new, transitional, or consolidating democracies. By strengthening legislatures, not only will aid dollars be used more efficiently, but also improved governance will ultimately reduce aid dependence and promote self-sufficiency.

Decentralization of the Aid Program, Budget Support, and the Trend to Country Ownership

An Overseas Development Institute report (Lawson and others 2002) evaluating the concept of general budget support as an aid instrument concluded that fragmented, ad hoc aid projects were failing to deliver results. It went further to say that, on a cumulative basis, these projects may have undermined aid agency development objectives for effective use of aid dollars as a result of the following:

• High transactions costs caused by the uncoordinated multiplicity of different reporting and accounting requirements

- Inefficient spending dictated by donor priorities and procurement policies
- Undermining of state systems by special staffing arrangements and parallel structures
- Corrosion of democratic accountability with mechanisms designed to satisfy donor rather than domestic constituencies
- Unsustainability of positive benefits beyond the short term because reliance on ongoing donor funding often undermined long-term sustainability
- Corruption, fraud, and rent seeking as frequent features of the management of projects that proved difficult to eliminate because of independence from government control

The concerns noted by the Overseas Development Institute reinforced a growing realization in the international donor community that building recipient government capacity and accountability to its own citizens for service delivery represented the most sustainable way of reducing poverty in the long term. This new thinking was explored in the OECD's Joint Evaluation of General Budget Support 1994–2004, a study commissioned by 24 aid agencies and seven partner governments. Undertaken by the University of Birmingham, the study resulted in multiple documents whose purpose was "to assess to what extent and under what circumstances general budget support … is relevant, efficient and effective for achieving sustainable impacts on poverty reduction and growth" (Dom 2007, 1). For the most part, the OECD study was positive about the potential for budget support as a means of providing aid and recommended that donors should assimilate partnership general budget support as part of their long-term assistance plans. At the same time the study emphasized the critical need for adequate public financial management systems in aid-receiving nations for the model to be a success.

A United Nations conference on financing for development held in Monterrey, Mexico, in 2002 resulted in a developing consensus that donor focus should shift more toward the effectiveness of aid rather than the dollar amounts given, while recognizing the latter also had to increase for the Millennium Development Goals on alleviation of poverty to be realized. Crucially, the conference cemented growing acceptance that lack of governance was a constraining factor for efficient development and that increased emphasis needed to be placed on strengthening and shaping the capacity of the domestic institutions of government in partner countries. Following on from Monterrey, the requirement for ensuring good governance as a means to improve development achievement led to a series of four key OECD forums that took place over the following six years with the goal of developing a globally endorsed set of principles for improving aid effectiveness: the High-Level Forum on Harmonization in Rome (2003), Paris High-Level Forum on Aid Effectiveness (2005), Accra High-Level Forum on Aid Effectiveness (2008), and Busan High-Level Forum (2011). The key themes, policies, and outcomes of the three most recent forums are summarized in the following subsections.

The Paris Declaration on Aid Effectiveness

Building on the principles from the 2003 Rome Forum (OECD 2003), the 2005 Paris High-Level Forum on Aid Effectiveness responded to an awareness that the international aid community needed to do much more to formally align donor and recipient country interests. The intent was to ensure that plans and objectives were harmonized to promote more sustainable development and self-sufficiency in partner countries through strengthened domestic institutions.

The Paris Declaration on Aid Effectiveness (OECD 2005) firmly established the importance of mutual accountability, endorsing five key principles—alignment, harmonization, managing for results, mutual accountability, and ownership—and detailing the specific commitments expected of each of the respective stakeholders in the aid process. As a result, the international aid domain was to take on a radically new approach whereby recipient countries would formulate their own national development strategies and donors would align their efforts to support those strategies. Importantly, Paris affirmed that development should be undertaken using the recipient country's systems so that donors "rely to the maximum extent possible on transparent partner government budget and accounting mechanisms" (OECD 2005, 5). For countries to take ownership of budget support and other means of aid channeling, the need to address lack of institutional capacity through strengthening of accountability institutions was also acknowledged. The donor community committed to providing resources for upgrading budget processes and strengthening parliaments to ensure that resource allocation was both efficient and democratic.

The Accra Agenda for Action

The Accra meeting sought to further improve coordination between donors and partner countries and to accelerate progress made since 2005 by focusing on three key themes (OECD 2008):

- First, the need for greater emphasis on ownership of development processes and formulation by the aid-receiving countries, and more use of their own systems to implement and evaluate these processes
- Second, the need for more inclusive partnerships, with full involvement of all stakeholders
- Third, the requirement for results of aid finance provision to be transparent, measurable, and accounted for

Reviewing the progress (or in some areas, lack of progress) made since the Paris forum, the follow-up meeting in Ghana in 2008 tackled issues of devolving responsibilities of budget allocation and control to aid-receiving nations. The proponents of devolution saw the opportunity for strengthening domestic accountability systems by emphasizing transparent budgetary mechanisms, implementing public financial management reform, and accounting for results. Detractors expressed their concerns that reliance on country systems risked buttressing governments' grip on development policies and funding at the expense of civil society actors. They also pointed to the danger of countries with weak administrative capacities focusing on being accountable to donors rather than to their own citizens. As a consequence, Accra was notable in further elevating the important role that parliaments must play in the development agenda.

The Busan Partnership

The fourth of the high-level forums convened in the Republic of Korea on December 1, 2011. It resulted in endorsement by 160 countries and 50 other organizations of the Busan Partnership for Effective Development Co-operation, announced by the OECD as a "a statement of consensus that a wide range of governments and organizations have expressed their support for, offering a framework for continued dialogue and efforts to enhance the effectiveness of development co-operation" (OECD 2011b, 2). This newly forged global partnership not only deepened the commitments of aid development stakeholders, but also set out formal standards and explicit actions for accelerating the implementation of the aid-effectiveness commitments agreed in Paris and Accra (OECD 2011a). Some examples are as follows:

- Use results frameworks and country-led joint assessment and coordination arrangements of country systems.
- Strengthen link between procurement and public financial management systems.
- Provide more development assistance to parliaments and create an enabling environment for civil society organizations to act as development actors.
- Implement a common standard of electronic publication of information, both timely and comprehensive.

In particular, the Busan Partnership emphasized how aid on its own will not break the poverty cycle but should be provided in conjunction with other development initiatives aimed at improving the lives of the general populations of developing countries. The Busan Partnership, therefore, takes a strong position on developing and mobilizing domestic resources in addition to the strengthening of national capacity of institutions and country-led directives.

Implications of the New Aid Model for Donor Organizations

So far this chapter has reviewed the evolving international aid landscape and, in particular, how the aid community has progressively developed and refined its thinking in relation to how long-term development goals can best be accomplished. It has recorded the trend away from project funding toward budget support, with recipient states assuming responsibility for determining how monies are spent. Moreover, to provide the confidence that funds are being invested effectively for the benefit of the population and in line with agreed development goals, donors are committed to providing assistance for strengthening institutions of governance—and parliament in particular—in the partner countries. This section looks in more detail at the changing way aid is being administered in the donor community, the changes that are in train, and the increasingly important role that parliaments play in ensuring that development objectives are met with optimum efficiency.

In the past 10 to 15 years, multilateral donors have adapted to the changing aid environment, in particular providing funding in the form of generalized budget support or sector budget support to countries that exhibit good governance. The World Bank has led the way in this realignment of aid policy, with the European Union and other donor organizations increasingly adjusting priorities in favor of this aid modality. Focusing on the World Bank as an example, this section looks at how a major international development organization has adapted its policies and procedures to this new approach to aid. More specifically, it shows how country ownership of financial assistance has gone hand in hand with allocation of resources by the World Bank to assist in strengthening the institutions of governance in partner countries to ensure that provided funds are managed properly and used to best effect. It also looks briefly at the implications of the new aid model for bilateral donors.

The World Bank's Changing Approach to Aid Delivery

Until quite recently, the World Bank was reluctant—even unable under its constitution—to engage in the sensitive issue of state governance, which was deemed to fall outside its remit because of its political connotation. Article IV, section 10, of the Bank's Articles of Agreement, formulated at the Bretton Woods Conference in 1944, states:[1]

> The Bank and its officers shall not interfere in the political affairs of any member; nor shall they be influenced in their decisions by the political character of the member or members concerned. Only economic considerations shall be relevant to their decisions, and these considerations shall be weighed impartially in order to achieve the purposes stated in Article I.

This position was reinterpreted in the World Bank's first publication on governance (World Bank 1991). In practice, the Bank's stance changed after research demonstrated an empirical link between governance and developmental success. Through the 1990s, World Bank strategists increasingly supported the idea that in-country governance is crucial to economic success and that, in its absence, the Bank's objectives to alleviate poverty would be unattainable because of poor economic management in developing countries. In concert with consultative discussions held by the International Anti-Corruption Conference, the World Bank issued a new governance and anticorruption (GAC) strategy (World Bank 2007) to incorporate this thinking into its objectives. This realigned strategy allows resources to be channeled to the areas of improving state accountability and anticorruption measures in partner countries without compromising the political interference clause of article IV.

The 2007 GAC strategy, plus its revision in 2012 (World Bank 2012), in effect constituted a fundamentally new paradigm for the Bank's strategic management.

Its intention was to assist in achieving transparent and competent state systems in partner countries by strengthening the institutions of government. This major shift in strategy was crucial in renewing the World Bank's mandate. By making its delivery of initiatives more effective and encouraging client countries to develop their own network of systems and institutional architecture, the Bank would further align its approach to long-term sustainable development goals.

The 2012 revision sought to redefine and reposition the GAC strategy in the light of recent world events, such as the global financial downturn and Arab Spring, both of which had dimensions of citizen-led calls for greater state accountability and transparency, further validating the Bank's new thinking. As Linda Van Gelder, World Bank director for public sector and governance put it: "Our updated strategy will help build those institutions for better governance around the world. The Bank's country-driven approach will be supported by global initiatives against anti-corruption and malfeasance" (Van Gelder 2012).

The Bank's Operational Policy 8.60, on development policy lending, affirms that

> The World Bank is committed to country-led policies and programs because it recognizes that reform can succeed only when the country itself has ownership of the process. Stakeholder participation in the policy process helps build ownership by involving a variety of groups in formulating the policy and thus engaging their interest in its implementation. (World Bank 2004, 5)

The GAC strategy stressed six central doctrines and areas of action for the Bank:

- Scaling up, whereby the Bank will attempt to systematically tackle issues of governance
- Supporting country institutions, in which the importance of constructing and maintaining domestic accountability institutions is realized[2]
- Focusing on results with appropriate indexes in place to measure progress and levels of institutional quality
- Exercising risk management
- Improving global governance by incorporating governance dimensions into programs
- Ensuring effective alignment of accountabilities with resources in the organization of aid delivery

Changes brought about by the GAC strategy—rooted in evidential links between governance, corruption, poverty reduction, and aid effectiveness—enshrine in World Bank policy the central tenet that economic efficiency will be achieved only in conjunction with strengthening domestic institutions in aided countries. This idea embraces the concept that open and accountable institutions are at the core of a well-functioning public sector and have a vital role in ensuring that the executive arm of government is deploying resources on behalf of the citizens to achieve the overall purposes intended. Crucially, as a result of the GAC strategy, the World Bank has moved away from exclusive dealings with the executive regarding international transfer of resources and toward using multiple entry points involving multiple stakeholders, including accountability institutions such as legislatures, civil society engagement, public financial management institutions, and the judiciary.

Ways Other Bilateral and Multilateral Donor Agencies Are Adopting the New Aid Model

Aid effectiveness is a priority for bilateral donors as well as for multilateral organizations such as the World Bank. In challenging times with ODA budgets under ever-increasing pressure, bilateral donors, too, are adapting to the changing economic and aid environments. They are increasingly persuaded by their own constituent stakeholders to ensure that aid dollars are deployed to maximum effect, with added pressure of scaling up aid to achieve the Millennium Development Goals.

Direct budget support channeled through country systems is widely seen as the optimum way to achieve efficiency, with the additional benefits of strengthening the recipient country's budget process and, in turn, improving administrative and fiscal capacities for long-term sustainability. The OECD's large-scale study, Joint Evaluation of General Budget Support 1994–2004, found that in using general budget support aid, "partner governments' transaction costs at implementation stage have been significantly reduced, by virtue of being able to follow standard government procedures rather than a multiplicity of donor ones" (Dom 2007, 3). Moreover, the old model of aid provision, where different donors would set up parallel organizations to perform essentially the same tasks, was not only wasteful in its use of funds, but also served to deepen developing countries' reliance on donors.

Numerous case studies involving both multilateral and bilateral donors show how the new aid model has proved to be more efficient and more universally accepted than the former approach to aid provision. For example, following serious flooding in 2000 in Mozambique, the reconstruction in-budget program proved far more effective at getting schools back up and running quickly than individual donor-sponsored systems initiated at the same time (Killen 2011). A World Bank study into the Ethiopian water sector in 2009 also observed that "parallel accounting systems being set up by centrally financed initiatives are inferior to the core integrated budget and expenditure management system" (World Bank 2009a, 70–71).

The European Commission initiated budget support pledges of more than €13 billion from 2003 to 2009, representing about a quarter of its aid commitments in the period, and 56 percent of these commitments were to Africa. Confirming that aid of this type provides "the strongest platform that we have to engage in a broad policy dialogue with our partner countries on key development issues," EuropeAid also acknowledged that where in the past project aid tackled the symptoms, budget support seeks to address the source of underdevelopment (European Commission 2011, 2).

The U.K. Department for International Development has also modified its approach to bilateral aid to address the "demand-side environment" on the basis that "funding is increasingly derived from domestic and international resource flows with increased oversight by respective parliaments and other bodies" (DFID 2013, 1).

Similarly, the U.S. Agency for International Development (USAID), accepting the principles developed from the high-level forums, has increased funding to host government institutions. In the 2012 financial year, 17.2 percent (including cash transfers) of its mission funds were given to local institutions, with half going to government bodies (national ministries) of partner countries. USAID is working with partner governments to increase the capacity of national institutions and processes so that when funding is provided, the agency has confidence that it is being managed effectively. By using tools to assess financial management capacity in partner countries, the agency is able to identify areas where potential may exist for misuse of U.S.-supplied funds.

A further example of the new aid model is the Tokyo Mutual Accountability Framework. Under the framework, donors pledged US$4 billion of civilian aid annually to the Afghan government with 50 percent in the form of budget support (Byrd 2013), thereby relying on the Afghan budget process and parliamentary oversight to control a substantial amount of aid income. This aid was issued on condition of improved governance and democracy during the leadership transition.

Despite these positive examples of aid agencies that have redirected funds to budget support and the associated backing for national management systems, many bilateral donors still harbor lack of trust in the national systems of many aid-dependent nations and their ability to exercise good governance. Hence, much bilateral aid continues "off budget," thereby denying local parliamentary oversight of its deployment. This confidence gap is counterproductive because it bypasses involvement of the very institutions that need to be strengthened to improve the long-term well-being of the country as a whole. As foreign direct investment and public-private partnerships assume greater importance than development aid in the overall mix of funding in developing economies, donors' willingness to work toward bolstering country institutions is more essential than ever to provide a governance legacy fit to meet the demands of increased economic activity and future capital inflows.

Budget Support and the Implications for Parliaments

As noted previously, budget support is progressively being adopted as the new modus operandi of aid assistance, in line with the policies and values agreed at the OECD forums on aid effectiveness. Donors—bilateral and others—increasingly favor budget support as being a more effective instrument for achieving the wide-scale development objectives encapsulated in the Millennium Development Goals. Moreover, it conforms to the new mandate for demand-led rather than supply-side aid delivery. Thus, whereas previously aid had been dedicated to large projects that invariably came with a plethora of donor-set strings attached, the trend now is toward aid provision based on a partnership approach, with responsibility for how funds are used assigned to the recipient partner.

More autonomy of the development agenda and more reliance on country systems to deploy development assistance and measure its effectiveness inevitably require robustness in the country's institutions. As the entity that is most attuned to public needs and preferences and that formally represents civil society, parliament is the most important and pivotal institution in legislating for and monitoring the budget and determining how aid money is spent. Parliament can oversee and improve the efficiency of public resource allocation, hold the executive to account, ensure that the executive fulfills its responsibilities in the interests of the citizens, approve budgets, collaborate with the auditor general to confirm value for money in ex post scrutiny of the budget, and challenge new legislation. Described by the European Commission as "the pre-eminent forum" for inclusive political dialogue and national debate, parliament is at the summit of the accountability tree (European Commission 2010, 112).

Sadly, in the past, this has not been the reality, and parliament has often been viewed as a weak link in the national budget process, but increasingly donors are recognizing its importance as the primary watchdog of their resources. Parliament is seen as occupying a unique position as the central hub of expertise and information on policy outcomes supplied by a number of accountability institutions. Constitutionally, if not always in practice, parliament has the authority to hold the executive to account on matters concerning public resources—including the effective use of ODA.

The World Bank's GAC strategy recognizes parliamentary capacity as a "co-equal branch of government" (World Bank 2009b, 20) whose legislative, oversight, and representation functions are the central powerhouse to delivering initiatives and ultimately reducing poverty. In fact, the very concept of mutual accountability between donors and recipients is underwritten by parliamentary programs that underscore its prerogative to "provide a check on the activity of government … providing government by explanation" (GOVNET 2012, 67). Parliament's ability to be transparent and work on behalf of the public through public accounts committees, research facilities, and auditors general creates the enabling environment for positive change, democracy, efficiency in implementation, and follow-up of accountability. "Institutional arrangements fundamentally affect public policy and the balance of power between political actors" (Wehner 2010, 18), and the strengthening of the parliamentary institution is accepted as the key to improving accountability of the executive in aided countries. Most important, ensuring that aid dollars are used efficiently and are properly incorporated into the national budget is in the interests of the well-being of the electorate as a whole. A World Bank guidance note on multistakeholder engagement, which was prepared as an outline of good practices for engaging multistakeholders in a consistent and flexible manner and to support the 2007 Implementation Plan for Strengthening World Bank Group Engagement on Governance and Anti-Corruption, astutely states, "Parliaments should thus be approached with deference" (World Bank 2009b, 20).

The GAC strategy and the World Bank's open budget initiative provide one example of how governments in developing countries are being helped to become more transparent in relation to managing national budgets. In particular, it recognizes parliament as being at the fulcrum of a fully accountable public financial management system and the entity with the most contact with demand-side civil society and most able to function as an effective monitoring mechanism. The World Bank's parliamentary support program is designed to help parliaments acquire the knowledge and skills required to perform their oversight tasks effectively. The World Bank, along with bilateral donors, has also developed the Public Expenditure and Financial Accountability (PEFA)[3] indicator system to monitor each stage of the national budget process. In a similar vein to the GAC strategy, the Demand for Good Governance agenda calls for an enhanced capacity of oversight functions in providing access to information, participation and planning, consultation, implementation, and follow-up to enable good governance.

Benchmarks Required for Mutual Accountability

In the organizational world, benchmarking is the practice of comparing the processes and performance metrics of one organization to best practices across the sector as a whole. In the field of parliamentary oversight of government, high-level measures are already in place—in particular the PEFA indicators that have been developed by a partnership of international interests, including the World Bank. They were originally intended as a means of formally assessing the public expenditure and financial accountability systems of each country to rate the country's capacities of governance, monitor progress over time, and make intercountry comparisons.

The PEFA indicators are useful as benchmarks of overall national performance assessment, monitoring, and comparison but are not sufficiently comprehensive or granular to allow detailed analysis of the functioning of individual procedures, departments, and mechanisms within the institutions of governance. In particular, the PEFA indicators are not suitable for designing, incentivizing, driving, and monitoring action plans for strengthening parliament.

To date, a set of universally recognized benchmarks for governance that can be used to design, implement, and monitor effective parliamentary strengthening initiatives to facilitate the mutual accountability that lies at the heart of the new aid modality has been absent both in practice and in the literature.

For the budget-support aid model to be rolled out further and with more assured success, the adoption of an agreed institutionwide approach to delivering parliamentary strengthening assistance in line with a set of recognized benchmarks of international best practice is seen as the next important step, because under the partnership and mutual accountability concepts developed at the OECD forums, both donor and receiver interests have compatible goals and objectives. To this end, having a universally accepted set of measures for parliamentary functioning and governance that provides for more efficient and streamlined consultation in agreeing on plans and roadmaps of action would make sense. Additionally, obtaining international consensus would be important in the development of the standards, which should be relevant and workable in the contexts of all types of legislatures, whether parliamentary, presidential, or semipresidential.

In the absence of such a consensus, a number of international organizations, including the World Bank, the United Nations Development Programme, the Commonwealth Parliamentary Association, the Inter-Parliamentary Union, and the Assemblée Parlementaire de la Francophonie (Parliamentary Assembly of La Francophonie), as well as nongovernmental organizations (such as the National Democratic Institute for International Affairs and the Parliamentary Centre) and the European Parliament, have formulated sets of benchmarks covering the functioning of democratic parliaments that establish accepted best practice and behavior standards for exercise by countries in receipt of budget-support assistance. Many of these benchmarks are contained in this book, and together they set the bar for comparison and evaluation. Although they do not constitute a universally agreed set of benchmarks and lack a common methodology, they demonstrate considerable overlap, duplication, and synergy between the different approaches and form a basis for setting specific goals for evaluating the effectiveness of donor-implemented parliamentary strengthening programs and enhanced infrastructural support mechanisms. They can be constructed to a highly detailed working level—as shown in the case studies contained in this book—covering minutiae of activities considered relevant to a properly functioning democratic parliament. These details include, for example, the nature, extent, and availability of research services to members and their staffs; technical capacity available to parliamentary committees, staff members, and members of parliament; and presence of audio-recording equipment in committees. In deploying these new detailed benchmarks, donors must, however, be cognizant of the care required to ensure that the accountability they are intended to enforce between donor and receiver does not distract or detract from the more important accountability that must exist between government institutions and the general public. In other words, as experienced by many public institutions across the world that have had target-based cultures imposed deep into their organizations, a danger exists of creating a checkbox mentality and "analysis paralysis" to the detriment of commonsense management and good public service.

That said, provided the design and use of the benchmarks are carried out sensibly and sensitively, their implementation will do two things. First, it will assure international donors of the capacity of parliament to perform its oversight role effectively and consequently that aid resources supplied as budget support will not be misused or squandered. Second, it will act as an instrument for incentivizing countries to accept legislative strengthening initiatives.

The donor community's interest in benchmarks is clear. We now need to determine the benefits of these measurements to parliaments themselves. What intrinsic benefits will parliaments gain from these benchmarks, apart from the obvious advantage that their adoption stands to encourage international aid givers to move to a budget support approach and so provide the country with independence and discretion over its use of aid monies? Keith Schulz (2010) of USAID reasons that the very existence of the benchmarks by itself will, almost subliminally, motivate parliamentarians to identify the shortfalls in their own institution by reference to the performance of other parliaments, which will automatically incentivize them to strive for continuous improvement. Therefore, benchmarks perform an important function as self-assessment tools that allow change to occur from within rather than being imposed from outside, thus enabling a parliament to monitor its own progress regarding legislative strengthening agendas.

More generally, the existence of a set of standards has the potential to promote greater accountability and transparency in the legislature and in the political system as a whole. The benchmarking process can be used to identify shortfalls and gaps between intended legislative powers and actual practice, thus monitoring parliamentary performance over time through, for example, data collected from surveys of members of parliament and staff members covering actual events and perceptions. A less tangible but equally important advantage of parliamentary benchmarks is the message that their use communicates to the electorate at large in terms of promoting transparency and accountability through increased dissemination of information and, in some cases of underperformance, naming and shaming. Finally, benchmarks are a useful component for training members of parliament, parliamentary staff members, civil society advocacy groups, and the wider public. Such training can be especially useful in nations with a limited history of democracy and transparency. In these countries, promoting an understanding of the key touch-points and performance expectations is helpful in the transformation from rubber-stamping institution to fully functioning, democratic oversight body. Equally, benchmarks can be used as a basis of educational aid for field officers and staff members of donor organizations who do not have formal backgrounds in legislative oversight or parliamentary programs. In these cases, a universal set of standards and objectives can be enormously beneficial as an aid to understanding processes and priorities entailed.

Conclusion

By adopting a methodological approach that started with the background and reasons for the changing pattern of international aid delivery and understanding its implications for the aid providers and for the legislatures in recipient countries, this chapter has identified and documented the need for universally recognized benchmarks for evaluating a parliament's capacity for national budget oversight. It does so for two reasons. First, the international donor community has neglected providing aid assistance for legislative strengthening in the past—a situation that is no longer tenable in light of the new aid modality that embraces country ownership and the concept of mutual accountability. And second, a logical line of reasoning is needed to explain why allocating funds and other resources to support and strengthen parliaments in aid-receiving countries is important for meeting long-term sustainability and poverty alleviation objectives. Moreover, such assistance falls within the donor community's mandate and is unequivocally in the interests of the developed world.

This chapter first charted how, at the approach of the millennium, there was growing debate about overseas aid effectiveness and legitimate concern that its capacity to relieve poverty had fallen well short of public and donor expectations. This realization led to the series of high-level OECD forums on aid effectiveness at which discussions took place to reconstruct the aid assistance platform using the principles of country ownership and strengthening of institutions in the recipient countries. The new thinking led to the prioritization of budget support as a mode of aid servicing, which, in turn, focused attention on the competencies of public financial management systems and domestic accountability institutions. This focus was essential for two principal reasons: first, for processing the increased inflows of ODA, and second, for ensuring that ODA was used efficiently to benefit the population as a whole and to address issues of poverty, employment, sustainability, and disadvantage as intended. The chapter then turned to a discussion of the implications of the new aid model for the parliaments in aid-receiving countries. Parliament constitutes the most important component of governance for ensuring that donor monies are used effectively and in line with the overall objectives of overseas aid, and so parliaments are arguably the institutions in most need of support and strengthening. The chapter then analyzed how the process of strengthening legislatures requires a means to evaluate and compare detailed aspects of parliamentary functioning. The chapter concluded with a discussion of the need for implementing a universal standardized set of parliamentary benchmarks and their implications for all stakeholders involved.

Notes

1. This quotation is from the Articles of Agreement of the International Bank for Reconstruction and Development, opened for signature December 27, 1945, 60 Stat. 1440, 2 U.N.T.S. 134, as amended February 19, 1989. The full text of the Articles of Agreement is available at http://web.worldbank.org/WBSITE/EXTERNAL /EXTABOUTUS/ORGANIZATION/BODEXT/0,,contentMDK:50004943~menuP K:64020045~pagePK:64020054~piPK:64020408~theSitePK:278036,00.html.

2. The need for transparency has been further accentuated by recent events in the Middle East and North Africa, highlighting the problems that arise when governance is not open and when the public link with government and scrutiny over issuance of public resources is weak. Key to sustainable development, the GAC strategy again reiterated the need to defer to country systems whenever and wherever possible with full engagement in the country's own agenda for its own development policies (country ownership).

3. The PEFA program was founded in December 2001 as a multidonor partnership between the World Bank, the European Commission, the United Kingdom's Department for International Development, the Swiss State Secretariat for Economic Affairs, the French Ministry of Foreign Affairs, the Royal Norwegian Ministry of Foreign Affairs, and the International Monetary Fund.

References

Byrd, William. 2013. "Travails of Mutual Accountability in Afghanistan." *Foreign Policy*, May 28. http://southasia.foreignpolicy.com/posts/2013/05/28/travails_of_mutual _accountability_in_afghanistan.

DFID (U.K. Department for International Development). 2013. "International Development Evaluation Policy, May 2013." DFID, London.

Dom, Catherine. 2007. "What Are the Effects of General Budget Support?" The Joint Evaluation of General Budget Support 1994–2004: Thematic Briefing Papers, UK Department for International Development, Glasgow.

European Commission. 2010. *Engaging and Supporting Parliaments Worldwide: Strategies and Methodologies for EC Action in Support to Parliaments*. Luxembourg City: Publications Office of the European Union.

————. 2011. "Background Information on Communications 'Agenda for Change' in EU Development Policy and EU Budget Support. Press Release, Brussels, October 13. http://europa.eu/rapid/press-release_MEMO-11-696_en.htm?locale=en.

GOVNET (Development Assistance Committee Network on Governance). 2011. "Draft Synthesis of Guidance and Key Policy Messages on Aid, Accountability, and Democratic Governance: Programme on Improving Support to Domestic Accountability." Presented at a meeting of GOVNET, Paris, June 6–8.

————. 2012. "Draft Orientations on Aid, Accountability and Democratic Governance." Presented at a meeting of the GOVNET, Paris, July 12. http://www.g20dwg.org /documents/pdf/view/296/.

Killen, Brenda. 2011. "How Much Does Aid Effectiveness Improve Development Outcomes? Lessons from Recent Practice." Busan Background Papers, Fourth High-Level Forum on Aid Effectiveness, Busan, November 29–December 1.

Lawson, Andrew, David Booth, Alan Harding, David Hoole, and Felix Naschold. 2002. *General Budget Support Evaluability Study: Phase 1: Final Synthesis Report*. London: Overseas Development Institute.

OECD (Organisation for Economic Co-operation and Development). 2003. "Rome Declaration on Harmonisation." First High-Level Forum on Aid Effectiveness, Rome.

————. 2005. "The Paris Declaration on Aid Effectiveness." Second High-Level Forum on Aid Effectiveness, Paris.

————. 2008. "Accra Agenda for Action." Third High-Level Forum on Aid Effectiveness, Accra, Ghana.

————. 2011a. "Busan Partnership for Effective Development Co-operation." Fourth High-Level Forum on Aid Effectiveness, Busan, Republic of Korea, November 29– December 1.

————. 2011b. "Busan Partnership for Effective Development Co-operation: Frequently Asked Questions." Fourth High-Level Forum on Aid Effectiveness, Busan, Republic of Korea, November 29–December 1.

Schulz, Keith. 2010. "Donor Reaction to Benchmarks and Standards: Improving Parliamentary Development Support." Presented at the International Conference on Benchmarking and Self-Assessment for Democratic Parliaments, Paris, March 4.

Van Gelder, Linda. 2012. "Governance and Anti-corruption Remain Integral to World Bank's Work." World Bank Press Release 2012/352/PREM, Washington, DC, March 27.

Wehner, Joachim. 2010. *Legislatures and the Budget Process: The Myth of Fiscal Control.* New York: Palgrave Macmillan.

World Bank. 1991. "Managing Development: The Governance Dimension." Discussion paper 34899, World Bank, Washington, DC.

———. 2004. "Supporting Participation in Development Operations." In *Good Practice Notes for Development Policy Lending.* Washington, DC: World Bank. http://siteresources .worldbank.org/INTCOUNTECONOMICS/Resources/GPNChapter5Participation .pdf.

———. 2007. *Strengthening World Bank Group Engagement on Governance and Anticorruption.* Washington, DC: World Bank. http://siteresources.worldbank.org /PUBLICSECTORANDGOVERNANCE/Resources/GACStrategyPaper.pdf.

———. 2009a. *Ethiopia Public Finance Review.* Report 50278-ET. Washington, DC: World Bank.

———. 2009b. "Guidance Note on Bank Multi-stakeholder Engagement." Document 49220, World Bank, Washington, DC.

———. 2012. "Strengthening Governance: Tackling Corruption—The World Bank Group's Updated Strategy and Implementation Plan." Document 67441. World Bank, Washington, DC.

CHAPTER 8

The Role of Parliamentary Monitoring Organizations

Andrew G. Mandelbaum and Daniel R. Swislow*

Introduction

Over the past decade, international parliamentary associations and their members have developed standards, benchmarks, self-assessment frameworks, and norms regarding the characteristics of a democratic parliament. During the same period, citizen-based groups have increasingly begun to recognize the importance of parliaments in consolidating democratic governance and to monitor the functioning of parliaments or their individual members. At present, more than 220 parliamentary monitoring organizations (PMOs) monitor more than 90 national parliaments worldwide. These organizations work to strengthen a number of components of democratic governance, including the accountability of parliaments to the electorate, citizen engagement in the legislative process, and access to information about parliaments and their work. They also show a growing capacity to encourage and support parliamentary reform. Many PMOs develop new technologies to facilitate the exploration of legislative information by citizens or to advance online collaboration and exchange between parliamentarians and their constituents.

Despite the common goal of PMOs and the international parliamentary community to strengthen the democratic functioning of parliaments, international discussions on these issues have mostly taken place until recently on separate, parallel tracks. Within the international parliamentary community, these conversations have led to the adoption of a variety of standards and self-evaluation frameworks for democratic parliaments, along with other guidelines on the release of parliamentary information and the use of technology to enhance parliamentary work and the engagement of citizens. The PMO community,

*This chapter was initially drafted in August 2013 when both authors worked as governance specialists at the National Democratic Institute (NDI). At the time of publishing, Andrew Mandelbaum is a co-founder of SimSim-Participation Citoyenne, a Moroccan parliamentary monitoring organization. Daniel Swislow is a senior partnerships officer and a governance specialist at NDI.

in contrast, only entered the debate on international standards for democratic parliaments in September 2012, with the launch of the Declaration on Parliamentary Openness.

The declaration calls on parliaments to increase their commitments to openness and citizen engagement through a concrete set of principles. The declaration, which takes into account nearly 130 supporting organizations from 75 countries, has received positive attention from the international parliamentary community. The PMO community has also received invitations to present the document to parliamentarians and parliamentary staff at international forums hosted by organizations such as the Inter-Parliamentary Union (IPU), the Global Centre for Information and Communication Technologies (ICT) in Parliament, and the Commonwealth Parliamentary Association (CPA). Recently, the Organization for Security and Co-operation in Europe Parliamentary Assembly (OSCE PA) endorsed the declaration (McKenzie 2013). PMOs have also begun to participate in broader discussions on government openness by way of (a) the Open Government Partnership (OGP), which is a multilateral initiative through which 60 governments are making commitments to work with civil society to become more open, engaging, and accountable, and (b) networks focused on freedom of information.[1]

Building on research conducted jointly by the National Democratic Institute for International Affairs (NDI) and the World Bank (Mandelbaum 2011) and informed by discussions taking place within the global PMO community, this chapter explores the potential for increased collaboration among PMOs and parliaments to stimulate positive effects on democratic parliamentary development. The main conclusions of this chapter are twofold. First, PMOs have demonstrated a capacity to add significant value to the development of standards frameworks for democratic parliaments. Second, collaboration between PMOs and the international parliamentary community can have mutually reinforcing benefits that strengthen representative democracy.

Roles and Effects of PMOs

In recent years, the role of civil society organizations in monitoring parliaments has increased dramatically throughout the world. Because PMOs operate at the nexus of civil societies, media, parliaments, and citizens, their approaches to parliamentary monitoring often vary. Some PMOs aggregate and analyze parliamentary information, presenting it in ways that are easier for the broader public to digest. Other PMOs create scorecards and indexes that use information about members of parliament (MPs) and their parties (for example, data on attendance, floor speeches, and votes) to evaluate their levels of activity in parliaments and, in some cases, in their constituencies. Many PMOs take a more qualitative approach to assessing parliaments and their institutional development, sometimes focusing on specific issues such as committee effectiveness or adherence to democratic principles (for example, inclusion and transparency). Moreover, some PMOs track and explain legislation to educate citizens and MPs about issues coming before parliaments, whereas others prioritize parliamentary accountability by tracking MPs' campaign pledges and asset declarations, party voting patterns, or even parliamentary adherence to rules of procedure.

The increasing focus of civil society on the functioning and performance of legislative institutions has expanded with the use of new technologies that are profoundly changing the way legislative information is used. Tools exist now that automatically aggregate publicly available information from parliamentary websites, databases, and other sources and then organizes the data into formats that are easy for citizens to understand, search, and analyze. An example of a website powered (in part) by such a tool is Scout, developed by the U.S.-based Sunlight Foundation. Scout allows real-time searches and alerts of references to keywords, phrases, or specific laws across multiple federal and state legislative and regulatory databases. It has also helped freedom of information advocates detect and defeat proposed exceptions to the Freedom of Information Act (Lee 2012). Informatics may also be used to create visualizations, such as infographics showing changes to a legislative text over time,[2] or to facilitate citizen engagement in the political process using techniques that allow citizens to comment on legislation or converse with MPs.

Although more research is needed, increasing evidence shows that PMOs can encourage accountability of parliaments to the electorate, facilitate citizen participation in parliamentary processes, and improve citizen access to information about parliaments and their work. In India, for example, an assessment of a campaign to create report cards of parliamentarians found that the project helped to decrease cash-based vote buying and to increase voter turnout, among other positive results (Banerjee and others 2011). A study of a parliamentary scorecard campaign in Uganda found that voters were sensitive to the information provided in the scorecards and that the scorecards attracted widespread media attention and were "hotly debated" by MPs (Humphreys and Weinstein 2012, 4). Although the study ultimately found little evidence that the scorecards caused citizens to change their votes, an Afrobarometer poll conducted closer to the 2011 elections indicated greater citizen awareness of the scorecards than was recognized by the study (Afrobarometer 2010). Despite substantial anecdotal evidence linking increases in attendance and participation by MPs (two basic indicators used to assess whether they change their behavior when being held accountable) to the scorecard, the study's authors acknowledge inability to detect a direct causal relationship between these indicators as a limitation of the study.

In many countries, a growing number of citizens resort to PMOs to learn about their parliaments' actions. During the first half of 2012, it is estimated that between 5 million and 10 million individuals accessed information from GovTrack, a PMO that monitors the U.S. Congress and its partners (Bruce and others 2012). In Colombia, a widget displaying information from the PMO Congreso Visible on the website of a major news outlet was used 65,000 times in a single day (Michener 2012b). PMOs have also become critical sources of information for journalists. For example, PRS Legislative Research in India has conducted trainings for more than 800 members of the media (Power and Shoot 2012, 54) to encourage effective and responsible use of the information that is accessed from their organization.

Although PMOs place a strong emphasis on using parliamentary information to shed light on the effectiveness of parliaments, in most cases their work is not limited to just monitoring. Evidence suggests that PMOs often take approaches that are successful in facilitating greater citizen engagement and in building more constructive relationships between parliaments and citizens. In Germany, a website that facilitates discussion between citizens and parliamentarians receives 350,000 unique visits per month, and 80 percent of the more than 100,000 questions that have been asked of parliamentarians through the platform have been answered. According to one parliamentarian interviewed by *Spiegel Online*, the website Abgeordnetenwatch.de (Parliament Watch) provides him with "one of the only chances to get to know the people in my constituency" (Glader 2012).

PMOs also often directly advocate parliaments for greater transparency and have affected policy change in many countries. In Argentina, a coalition of PMOs has signed a memorandum of understanding with the president of the National Congress that allows the PMOs to participate in regular working group meetings to help improve the transparency record of the National Congress (Swislow 2012). In Brazil, members of Transparência Hacker helped an MP develop a new Freedom of Information bill that provided a series of recommendations that were included in the final law (Michener 2012a). Where efforts at collaboration with parliaments have failed, PMOs in the Kyrgyz Republic, Romania, Tunisia, and many other countries have demonstrated their ability to affect policy by suing parliaments to ensure compliance with rules governing access to parliamentary information.

Development of Normative Frameworks for Democratic Parliaments and Global Emergence of PMOs

Since 2006, democratic norms and standards, benchmarks, and self-assessment tools have been developed and approved by several of the largest international parliamentary associations, including the IPU, the CPA, and the Assemblée Parlementaire de la Francophonie (Parliamentary Assembly of La Francophonie, or APF). More recently, the Southern African Development Community Parliamentary Forum (SADC PF) and the Parliamentary Confederation of the Americas (COPA) have developed standards and are now field-testing associated self-assessment frameworks.[3] International discussions have shifted to focus on areas of consensus (and nonconsensus) among the normative frameworks and on sharing experiences from the emerging body of good practice on the application of parliamentary self-assessment tools (WBI and UNDP 2010).

Although PMOs are developing innovative monitoring techniques and conducting substantive research and analysis of parliamentary functioning and performance, they have had limited experience with normative frameworks for democratic parliaments created by international parliamentary organizations. According to a survey of global PMOs that was published in 2011 by NDI and the World Bank, only 25 percent of respondents had developed evaluations using methodologies designed by international organizations (Mandelbaum 2011). Of these respondents, several mentioned familiarity with NDI's (2007) discussion document on democratic standards and the IPU's (2008) Self-Assessment Toolkit for Parliaments. Several other respondents indicated familiarity with Transparency International's National Integrity System Assessments and the International Institute for Democracy and Electoral Assistance's State of Democracy Assessment Methodology, both of which focus on governance issues more broadly but include assessment areas specific to parliaments. For a case study on a PMO's successful use of benchmarks, see box 8.1.

Until recently, debates among international parliamentary associations on normative frameworks for democratic parliaments have generally not benefited from the systematic engagement of PMOs. This lack may be, in part, because discussions on normative frameworks have generally taken place in venues typically reserved for MPs with the purpose of building parliamentary buy-in for the resulting frameworks. However, the lack of engagement of PMOs can also be attributed to their limited presence in the international arena and relatively recent emergence as a global community of practice. Most PMOs operate in specific country contexts and, up until recent years, few have benefited from opportunities to network or collaborate within an international context. With the exception of the establishment of a Latin American regional network and

---

Box 8.1 Parliamentary Monitoring Organizations' Use of Benchmarks: PILDAT

Emerging practice demonstrates the possibilities for more positive collaborations between parliamentary monitoring organizations and parliaments. The Pakistan Institute of Legislative Development and Transparency (PILDAT) conducted an evaluation of the Pakistani National Assembly in cooperation with members of parliament, analysts, and members of the media using the framework of the IPU's (2008) Self-Assessment Toolkit for Parliaments (see Mandelbaum 2011, 56). The 28 participants (half of whom were members of parliament) were asked to rate the National Assembly by answering questions from this six-section toolkit. The final report (PILDAT 2009) states the results and provides recommendations developed by participants to improve the parliament's effectiveness.

Although many of the recommendations have yet to be implemented, the secretary of the National Assembly credits the evaluation with prompting the decision to allow an opposition leader to chair the Public Accounts Committee and with encouraging the National Assembly's continued efforts at self-assessment. Commenting on the National Assembly's recent adoption of a private member bill to establish an internal research organization, PILDAT Joint Director Aasiya Riaz stated that it "took us years to sensitize MPs that this is something they need to undertake in their work. It's still in the teething stage, but an Act of Parliament has been passed" (Mandelbaum 2011, 56).

---

a handful of conferences focusing on the development of civic technologies (which included members of the PMO community but were more broadly focused), few opportunities were available to PMOs to share experiences and exchange good practices, thus limiting their ability to collaborate with one another and with parliaments internationally.

International PMO networking gained momentum in April–May 2012, when PMO leaders from 38 countries gathered in Washington, DC, for the first international conference that was exclusively focused on sharing information within the parliamentary monitoring community.[4] Hosted by NDI, the Sunlight Foundation, and the Latin American Network for Legislative Transparency (LALT Network), the conference sought to strengthen the ability of PMOs to advocate for increased access to parliamentary information, a challenge noted as the largest concern facing PMOs by the NDI–World Bank study.[5] Among the conference outcomes were the creation of a PMO network listserv to facilitate online communications, the creation of the OpeningParliament.org website and blog to serve as a channel for continued cooperation, and efforts by a number of participants to work together to help establish a regional PMO network in Africa.

The Declaration as a Contribution to International Parliamentary Norms and Standards

Another result of the Washington, DC, conference was a commitment by participants to develop—and support the development of—the Declaration on Parliamentary Openness. The process for developing the declaration is note-worthy in how it engaged the perspectives of the broad PMO community and in how it used new technologies. After the conference, a draft declaration that drew on conference discussions became available for public comment on the PublicMarkup.org website, which was designed by the Sunlight Foundation. It was also made available in downloadable formats, including the .ODT (open document) format. All comments from members of the global PMO community were posted on PublicMarkup.org to allow for discussion among interested individuals. In addition to the more than 70 PMOs that contributed to the drafting process, a number of parliamentary staff members and other representatives of the parliamentary community also participated. Moreover, the declaration was introduced to additional members of the PMO community, and academics focused on parliamentary information at the Open Legislative Data Conference in Paris, which was cohosted by the French PMO Regards Citoyens, the Center for European Studies at Sciences Po, and Medialab Sciences Po, resulting in further refinements.

Representing the first entry into the conversation on normative standards for democratic parliaments by PMOs, the declaration outlines norms aimed at enhancing parliamentary openness, transparency, and citizen participation in parliamentary work. It also highlights the role of parliament in ensuring citizen access and the reuse of parliamentary information and of government information more broadly. According to its introductory section, the declaration is "intended not only as a call to action, but also as a basis for dialogue between parliaments and PMOs to advance government and parliamentary openness, and to ensure that this openness leads to greater citizen engagement, more responsive representative institutions and, ultimately, a more democratic society."[6] In this vein, the declaration draws on standards frameworks and other publications developed by the international parliamentary community, as well as on good practices exhibited by parliaments themselves, to demonstrate the basis for each of its 44 provisions in both democratic practice and international norms.[7]

The declaration highlights the importance of technology for the functioning of democratic institutions worldwide.[8] The declaration's preamble notes that "the onset of the digital era has altered fundamentally the context for public usage of parliamentary information and the expectations of citizens for good governance," and that "emerging technology is empowering analysis and reuse of parliamentary information with enormous promise to build shared knowledge and inform representative democracy."[9] Although many of the international standards frameworks have had only a limited focus on the effect of technology on parliamentary functioning, the attention paid to this issue continues to grow within the parliamentary community. The final section of the declaration focuses specifically on online and digital communication of parliamentary information, building on documents such as the IPU's *Guidelines for Parliamentary Websites* and the IPU and United Nation Development Programme's *Global Parliamentary Report* (IPU 2009; Power and Shoot 2012), as well as on information about parliamentary use of ICT from the Global Centre for ICT in Parliament's *World e-Parliament Report 2012* (Global Centre for ICT in Parliament 2012).

The declaration, which is supported by nearly 130 PMOs in 75 countries, has forged new opportunities for collaboration with the parliamentary community and has provided a vehicle for speaking with a common agenda. The declaration was launched in September 2012 at the World e-Parliament Conference at the Italian Chamber of Deputies in Rome and was hosted by the United Nations and IPU through the Global Centre for ICT in Parliament. Since then, numerous international organizations of governments and parliaments have recognized the declaration's contribution to the creation of standards for democratic parliaments. The declaration was included on the agenda at the IPU's 127th Assembly in Quebec in October 2012. In May 2013, the CPA convened a study group that brought together MPs and PMO representatives to review the CPA's (2006) Recommended Benchmarks for Democratic Legislatures. It ultimately recommended adoption of the declaration, as well as further discussion on benchmarks for individual behavior of MPs. The study group also suggested open data principles so that parliaments could provide information in formats that could be easily processed using technology (CPA 2013). In July 2013, OSCE PA endorsed the declaration.[10] Additionally, PMOs and parliaments will come together to establish a working group on legislative openness as part of OGP, with the aim of expanding the scope of OGP country commitments to open government toward considering the legislative process.

The declaration has also had an influence at the national and subnational level. In Mexico, for example, efforts by PMOs to advocate for greater openness in the

Mexican legislature resulted in an endorsement of the declaration by the Mexican Senate, the passage of new internal rules on the release of parliamentary information, and the establishment of a working relationship between local PMOs and a senate committee focused on legislative transparency (Massó 2013). In the Czech Republic, PMOs have used the declaration to guide discussions with parliamentary staff members about increasing access to information. These discussions led to the release of parliamentary voting data in open and structured formats (Mráček 2012). The declaration has also been endorsed and used by state and local legislatures, including those of Buenos Aires, Argentina, and Andalusia, Spain.

Content of the Declaration

The declaration contains four sections comprising 44 provisions. Each section is described here, along with a brief discussion of how each one tracks with the normative frameworks for democratic parliaments.

First Section: Promoting a Culture of Openness

The declaration's first section underscores the importance of a parliament's creation of a culture of openness—both within a parliament and within a society at large—based on the principle that parliamentary information belongs to the public and should be reusable and republishable by citizens at their own discretion. Hence, the declaration calls on parliaments to use their legislative powers to enact citizens' right to parliamentary information and to harness their oversight powers to protect this right. The section also details the responsibility of parliaments to ensure inclusive citizen participation and a free civil society, to enable effective parliamentary monitoring, and to promote citizen understanding of parliamentary functioning. In fulfilling these principles, parliaments should work with PMOs and citizens to ensure that any information that is provided is complete, accurate, and timely.

The declaration's initial section draws strongly on standards frameworks created by international parliamentary bodies. For instance, COPA's (2011) Benchmarks for the Parliaments of the Americas requires that parliaments "recognize access to information as a fundamental right of citizens" (benchmark 4.1.1.1) and "foster a spirit of tolerance and promote all aspects of democratic culture" (benchmark 4.2.1.1). The CPA's (2006) benchmarks stipulate that opportunities be given for "public input" into the legislative process (benchmark 6.3.1), that "matters under consideration by the [parliament]" be made public in a "timely manner" (benchmark 6.3.2), and that parliament "promote the public's understanding" of its work (benchmark 9.1.4).[11] The *World e-Parliament Report*, acknowledging that a culture of openness and transparency may not be the predominant tradition, underscores that "a culture of transparency is consistent with the responsibilities of parliaments as the peoples' representatives, and it is consistent with the values of the citizens who live in the information society." It emphasizes that the basic principle of this culture "is that all information and documents should be made available and that exceptions should be established on a case by case basis" (Global Centre for ICT in Parliament 2010, 204).

Second Section: Making Parliamentary Information Transparent

The second section of the declaration expounds on the responsibility of parliaments to publish information about their activities and details the categories of information that should be available. These categories include information about a parliament's roles and functions, as well as information generated throughout the legislative process, such as draft legislation, agendas of parliamentary or committee activities, voting records, and transcripts of committee hearings and plenary proceedings. The section also calls on parliament to provide citizens with information on the administration of parliament, on its staff and budget, and on members of parliaments, including details related to issues of ethics and conflicts of interest (see box 8.2).

These provisions reinforce other standards for democratic parliaments and provide additional guidance to parliaments regarding the information that is to be made available to citizens. Concerning the publication of records of plenary proceedings, for instance, the CPA benchmark 2.7.1 provides that parliaments shall "maintain and publish readily accessible records of its proceedings" without specificity to the type of record or medium of publishing. Provision 21 of the

---

Box 8.2 Activities of Parliamentary Monitoring Organizations: Parliamentary Ethics

International benchmarks on democratic parliaments universally include provisions for ensuring ethical governance. Drawing on this work, the Declaration on Parliamentary Openness contains specific provisions on asset disclosure and member integrity, imploring in provision 24 that parliament "shall make available sufficient information to allow citizens to make informed judgments regarding the integrity and probity of individual members."

Where codes of conduct have been adopted or disclosure of the financial assets of members of parliament is required by law, parliamentary monitoring organizations (PMOs) often monitor members' compliance with the adopted measures. Such monitoring may amount to counting the members whose financial disclosure forms are provided on the parliament's website or identifying those who have breached specific norms enumerated in the code of conduct.

Whether or not a code of conduct has been ratified, PMOs have developed innovative approaches for monitoring parliamentary ethics. One approach is to conduct campaigns timed to coincide with parliamentary elections that seek to promote compliance with ethics rules and practices. The Al-Quds Center for Political Studies, for instance, used Jordan's 2010 elections to secure candidate signatures on an "Agreement with Jordan" that included a pledge to develop a parliamentary code of conduct once a candidate was elected. In other instances, PMOs may combine financial disclosure information with other data sources, such as public contract financing, to prevent potential conflicts and abuse.

Benchmarking and Self-Assessment for Parliaments • http://dx.doi.org/10.1596/978-1-4648-0327-7

declaration calls on parliament to "create, maintain, and publish readily accessible records of its plenary proceedings, preferably in the form of audio or video recordings, hosted online in a permanent location, as well as in the form of a written transcript or Hansard."

This section of the declaration also intersects extensively with the IPU's (2009) "Guidelines for Parliamentary Websites," although it does not limit the publication of parliamentary information solely to the web. For example, the IPU's guidelines call on parliaments to provide information on the roles and function of parliaments, including an "overview of the composition and functions of the national parliament and its constituent bodies" (section 1.3.a), the "budget and staffing of parliament" (section 1.3.b), and items such as "contact information for each member of parliament including his or her email address" (section 1.6.c). These provisions are mirrored in the declaration along with other provisions related to basic parliamentary information. Although broad normative frameworks detailing the responsibilities of a democratic parliament have left out much of the specifics on issues of transparency and accountability, recent contributions to the discussion, such as the declaration and the IPU's guidelines, have begun to fill in the gap.

Third Section: Easing Access to Parliamentary Information

The third section of the declaration seeks to ensure that parliamentary information is easily accessible to citizens. It includes the principle that multiple channels for distributing parliamentary information are necessary for enabling broad participation in parliamentary processes. The declaration also addresses other issues related to accessibility, including the use of plain language, the facilitation of access to information in different geographic parts of the country, and the availability of that data free of cost. Whereas standards frameworks provide that parliament must publish records, little attention is given to the media in which they should be published and made available to citizens. In provision 27, the declaration addresses this issue by specifying multiple methods for accessing parliamentary information, including "first-person observation, print media, radio and television broadcasts, and Internet and mobile device technology."

Standards frameworks for democratic parliaments universally include sections on accessibility of the parliament to citizens, the media, and civil society. For example, benchmark 2.1.1 of SADC PF (2010) states, "Parliament shall be accessible and open to citizens, civil society organisations, and the media, subject only to demonstrable public safety and work requirements." CPA benchmark 9.1.2 requires, for instance, that the media be "given appropriate access to the proceedings ... without compromising the proper functioning of the [parliament] and its rules of procedure." These standards documents also address other issues of the accessibility of parliamentary information, including the stipulation that parliaments must allow citizens physical access to their proceedings and that parliaments must accommodate the different language requirements of the citizens of their countries—standards that are also reinforced by the declaration.

Fourth Section: Enabling Electronic Communication of Parliamentary Information

The capacity of citizens to organize, analyze, parse, visualize, and otherwise reuse and republish parliamentary information depends on how parliaments present online information to citizens. The declaration's final section emphasizes the need for parliaments to release information in open and structured formats that are machine readable, such as structured XML (Extensible Markup Language).[12] Parliaments should take additional measures to ensure the usability of parliamentary information, such as giving preference to the use of nonproprietary formats, providing bulk downloads of parliamentary information, and using persistent URLs (uniform resource locators)[13] so that links to resources on the parliament's website remain constant over time. In addition, the declaration calls on parliaments to maintain and regularly update parliamentary websites, enable two-way communication with citizens, and develop effective search mechanisms for parliamentary information.

Standards frameworks adopted by international parliamentary associations have been largely silent on ICT. For example, benchmarks adopted by COPA, which are among the most recent, contain only limited provisions on ICT, including the stipulation that parliaments must "promote new information and communication technology and seek out ways in which technological advances could reinforce the democratic process and improve individual participation and decision making" (COPA, benchmark 4.1.1.3). Previous benchmarks were even more silent on topics such as providing information in electronic formats. Calling on parliaments to publish records "in a standard and consistent format that is appropriate and sustainable," the CPA's benchmarks for Caribbean parliaments scratch the surface of the standards for releasing parliamentary information (see CPA 2011, benchmark 2.7.1).

Notwithstanding the importance of technological change for the way parliaments function, parliamentary associations have addressed the issue largely through discussions that are somewhat distinct from those of standards frameworks. For example, the IPU released "Guidelines for Parliamentary Websites" (IPU 2009) and "Social Media Guidelines for Parliaments" (Williamson 2013) to address benchmarks for parliaments in the interest of reaching out to citizens via the web. The Global Centre for ICT in Parliament has also released several iterations of the *World e-Parliament Report*, which surveys a majority of the world's parliaments on technology issues and provides recommendations and benchmarks on issues of ICT (see, for example, Global Centre for ICT in Parliament 2012).

In justifying the section on electronic information, the declaration draws heavily on the work by the IPU and the Global Centre, as well as on the work of individual parliaments and other government resources. These justifications are cited extensively in the "Provision Commentary" on the declaration. For example, in discussing the declaration's provision calling on parliaments to provide information in an open and structured format, the document cites standards described in the Global Centre's report that "are needed to provide the functionality and flexibility required by parliaments for diverse requirements such as searching, exchanging, integrating, rendering, and particularly for ensuring the long term availability of digital records at an affordable cost. XML supports the values of transparency, accessibility, and accountability in a variety of ways."[14] The commentary further refers to parliamentary development in the European Union and in countries such as Chile and Brazil. PMO contributions to normative frameworks through the declaration are an important step forward as benchmarks surrounding the use of technology by parliaments continue to evolve and solidify. As described in the next section, collaboration between parliaments and PMOs on these issues provides substantial opportunity for advances in this area.

Benefits of Greater Collaboration between PMOs and Parliaments on Normative Frameworks

Given their shared interest in strengthening parliamentary performance, PMOs and parliaments alike could benefit from increased collaboration on developing normative frameworks for democratic parliaments. From the PMO perspective, engaging parliaments internationally in discussions of normative frameworks helps to legitimize civil society's right to monitor parliamentary work and to have input into the legislative process. From the parliamentary perspective, constructive dialogue with PMOs on norms and standards can foster greater citizen understanding of these norms and increase support for their implementation. As noted in the participant statement from the 2010 International Conference on Benchmarking and Self-Assessment for Democratic Parliaments, parliaments may also wish to engage with PMOs to improve the methodologies used by PMOs to better enable them to "engage in fair, responsible monitoring of parliamentary performance in accordance with international norms" (WBI and UNDP 2010, 5).

In particular, increased dialogue between PMOs and parliaments on normative frameworks can have three benefits: building and reinforcing constructive working relationships between PMOs and parliaments, developing comprehensive approaches for parliamentary monitoring and assessment, and combating public cynicism about parliaments.

Building and Reinforcing Constructive Working Relationships between PMOs and Parliaments

The relationship between parliaments and PMOs (and civil society more broadly) can often become unnecessarily confrontational. Although PMOs generally aim to strengthen parliamentary performance, some embark on a narrower approach that focuses on the conduct of MPs and exposure of the poorest performing members. Although these activities may achieve widespread media attention, they may provide disincentives for parliamentary collaboration to reform the underlying causes of poor parliamentary performance. Some MPs, particularly those who are shown to be among the poorest performers, may question the loyalties of these PMOs and accuse them of having political motives. PMOs that resort to confrontational approaches often do so in response to perceived resistance to monitoring activities by MPs (Mandelbaum 2011). Many PMOs suggest that despite their best efforts, parliaments are disinterested in civil society's monitoring of parliament until the findings attract attention in the media. The friction arising from this relationship may serve to reinforce broader tensions between parliaments and civil society.

As the CPA study groups and others have begun to demonstrate, discussions around normative frameworks for democratic parliaments can provide a constructive setting for PMOs and parliaments to explore these challenges. For PMOs, participating in this process allows them to engage MPs directly on issues related to the democratic performance of parliament and parliamentary assessment while also challenging MPs to understand the limitations PMOs face when trying to develop effective monitoring tools (particularly when confronted with a lack of information). Conducting assessments on the basis of normative frameworks for democratic parliaments—especially those developed by MPs or parliamentary associations—may provide PMOs with a more legitimate footing to assess parliamentary functioning and to build support for parliamentary reform.

For parliaments, greater engagement of PMOs on the international level and the development of tools to facilitate the sharing of information and best practices among them creates an opportunity to encourage fair and responsible monitoring by PMOs. As noted in the participants' statement from the International Conference on Benchmarking and Self-Assessment for Democratic Parliaments (WBI and UNDP 2010), engaging PMOs in discussions on democratic standards would elevate the issue of fair and responsible monitoring and improve the quality of civil society monitoring. Discussions of normative frameworks could help PMOs recognize weaknesses in existing assessment methodologies and challenge them to use methodologies that are most likely to stimulate reform rather than fuel public cynicism about their representative institutions.

Ultimately, international discussions on normative frameworks for democratic parliaments provide a common ground for PMOs and parliaments to explore the barriers to more effective collaboration. They may also permit the development of a common agenda for strengthening parliaments that can have a positive effect on parliament-PMO relations on a global scale.

Developing Comprehensive Approaches for Parliamentary Monitoring and Assessment

PMO methodologies and the normative frameworks stand to benefit from increased interaction between PMOs and parliaments. Some PMOs focus monitoring on the performance of individual MPs rather than on parliaments as institutions. This approach is fueled, in part, by the attractiveness of quantitative evaluations of individual performance to the media and citizens, particularly when they confirm perceptions of wastefulness and corruption on the part of individual MPs. PMOs also tend to focus their monitoring on the information that is available rather than on the information that would best help citizens to understand the work of parliaments or inform parliamentary reform. The structure provided by normative frameworks can help PMOs (a) use quantitative data to focus on measuring parliamentary development issues more broadly and (b) encourage the adoption of qualitative methods that facilitate advocacy efforts for democratic reform. New monitoring technologies are especially well suited to aggregating and visualizing large amounts of data, which makes an individual MP a useful unit of analysis for efforts to comprehend issues related to broader institutional functioning or behavior.

In contrast, the movement to develop normative frameworks for democratic parliaments has—in its initial stage—focused on forging broad consensus on a comprehensive listing of the normative characteristics of democratic parliaments. As a result, the process has primarily focused on the general qualities of parliaments as institutions, while paying less attention to the actions and behaviors of individual MPs. Now that several normative frameworks have been developed by parliamentary associations, there are opportunities not only to revisit more general norms to make them more measurable and concrete, but perhaps also to focus more on the norms related to individual parliamentary behavior.

Given the complementarity of the strengths and weaknesses of PMO assessment tools and parliamentary normative frameworks, parliamentary associations may serve as effective partners as PMOs seek to improve their tools. Parliaments may help PMOs to develop more holistic monitoring tools and to consider how these tools may be leveraged domestically to work with parliaments toward democratic reform. Conversely, PMOs may convince parliaments that because the actions and behaviors of MPs shape the characteristics that define the parliament and its functioning, increased emphasis on international norms for individual MP conduct could prove valuable. Collaboration between PMOs and parliaments can generate more robust mechanisms for measuring the degree to which standards are met by parliaments than by either community alone.

Combating Public Cynicism about Parliaments

Citizens are often skeptical of their parliaments, viewing them as aloof, corrupt, unresponsive, or ineffective. Furthermore, public approval ratings of parliaments are, in many cases, near historic lows. Although this public frustration may sometimes be warranted, it could be better channeled in ways that strengthen accountability structures and citizen engagement of parliament. When PMO activities reinforce public cynicism of the institutions that represent them, they may undermine democratic governance by bolstering the executive as an alternative to a corrupt or unproductive parliament. Mónica Pachón, director of Congreso Visible (Colombia), explains the problem as follows:

> Parliaments are not exactly popular—citizens don't look forward to [learning about] them.… If the discourse of the organization is similar to what the feeling of the people is—and doesn't question the negative image that people have about the Congress—then we are not doing much. We're saying "it's not worth it to inform yourself because there is corruption and clientelism and other things."… If citizens don't realize that Congress is a very important branch for a political system to work, then we're not going to be a democracy. (Quoted in Mandelbaum 2011, 21)

Benchmarking and Self-Assessment for Parliaments • http://dx.doi.org/10.1596/978-1-4648-0327-7

As discussed previously, many PMOs have sought to take collaborative approaches to parliamentary engagement to help address the causes of public cynicism. For instance, prior to releasing negative monitoring information, some PMOs have involved parliaments in discussions about the information and about ideas for reform that could help ensure a more constructive and engaging response by parliament. Similarly, many PMOs have shown promise in their efforts to facilitate citizen involvement in the legislative process. PMOs and parliaments can work more closely to help focus citizens' attention on parliamentary oversight activities—in addition to parliamentary treatment of the legislative process—where these activities promote good governance and strengthen the parliamentary institution vis-à-vis the executive.

Civic education is also an area for PMOs to work constructively with parliaments. By using innovative technologies and face-to-face meetings (such as back-to-school days for MPs) between MPs and citizens, there are opportunities to strengthen public understanding of the legislative institution. International discussions on normative frameworks between PMOs and MPs may also lead to shared approaches and strategies for educating citizens about their parliaments and for encouraging citizen input into the legislative process.

Harnessing Technology to Support Common Democratic Aspirations

Parliaments and PMOs are recognizing the potential for new technologies not only to enhance the relationship between parliaments and citizens, but also to enrich the policy process and parliamentary oversight. In some instances, technology has enabled parliaments to reduce costs[15] and enhance communications with MPs (Global Centre for ICT in Parliament 2010, 198). The creation of a legal documentation standard, Akoma Ntoso,[16] and an open-source software system designed specifically for parliaments, Bungeni,[17] have increased parliamentary access to tools that would allow for public provision of information in open and structured formats. Yet adoption of these tools remains slow. The *World e-Parliament Report* finds that just one-quarter of all parliaments are using XML for proposed legislation (Global Centre for ICT in Parliament 2010).

Where parliaments are slow to innovate, PMOs are rapidly advancing new technologies for improving citizen access to parliamentary information and to their MPs. Many PMOs share the code through open-source platforms and share their lessons learned through a variety of networking tools. Some are developing shared tools for presentation and analysis of parliamentary information that are intended to have regional or global application.[18] The collective knowledge within the PMO community enables some PMOs to contribute advice to parliaments on the adoption of new tools and technologies, as well as on the implementation of tools that would enhance parliamentary inclusivity and openness, qualities that are often invoked in democratic standards frameworks. Additionally, the experience of PMOs as both users and providers of parliamentary information gives them valuable insights on how to present parliamentary information in ways that can capture the interest of the broader public. Many are engaged in broader issues of open government and open parliaments and can provide contributions and advice to parliaments that are working to craft legislation that ensures that governments respect the right of citizens to open and transparent government information.

Strengthening the Capacity of PMOs to Refine and Apply Democratic Norms and Standards

Stakeholders and the international community can take a number of actions to strengthen PMOs and their collaboration with parliaments:

- *Make medium- to long-term investments in PMOs.* Few sustainable funding models have been identified thus far that apply to PMOs on a broad scale. Consequently, the international donor community continues to serve as an important funding source for PMOs, particularly those that serve developing countries. Although there is a tendency to fund parliamentary monitoring activities on a short-term basis, the pressure to show results may drive PMOs to reveal parliamentary malfeasance and to maximize press coverage to justify continued funding rather than focus on long-term challenges to parliamentary reform. Medium- to long-term support can provide PMOs the time that is necessary to develop effective approaches and methodologies and to forge credible working relationships with MPs. In some instances, the provision of funding to sustain an organization between election periods allows a PMO to plan over the life of the parliament and provide a more realistic window for producing results.

- *Support regional networking and peer-to-peer sharing among PMOs.* There are limited mechanisms for sharing good practices among PMOs at a regional level, despite the wealth of creative ideas that the PMO community has generated. The LALT Network has demonstrated the value of collaboration around a regional index for parliamentary transparency, thereby proving that initiatives driven by leading PMOs within a region can generate a cumulative effect that is greater than the sum of the individual members. An African PMO network has recently been launched, and there is strong demand in other regions for conducting collaborative initiatives or for establishing regional networks to meet strategic objectives. Regional networking offers a valuable opportunity for PMOs to share best practices and to harness their aggregate capabilities to improve the democratic functioning of parliaments. Moreover, individual PMOs can have substantial knowledge about the policies and good practices developed by the parliaments they monitor. Facilitating global and regional collaborations among PMOs can disseminate this knowledge regionally and help to improve the quality of PMO reform recommendations. The outcomes of such collaborations can also serve as a useful resource for parliaments.

- *Encourage regular engagement between PMOs and parliaments at the national and international levels.* Too often, apprehension and mistrust have characterized relationships between PMOs and parliaments. However, many PMOs and parliaments are finding that collaboration can have mutually beneficial outcomes, particularly with respect to implementing technology and improving citizen participation. Other benefits—in areas such as oversight—may also be realized. The Declaration on Parliamentary Openness, the standards frameworks, and documents such as the "Guidelines for Parliamentary Websites" (IPU 2009) and the *World e-Parliament Report 2012* (Global Centre for ICT in Parliament 2012) can serve as a starting point for constructive dialogue between PMOs and parliaments. The OGP Working Group on Legislative Openness, which officially launched at the OGP Annual Summit in late October 2013, may provide a useful forum for collaborative discussion among PMOs and parliaments.

- *Open parliamentary information and engage the PMO community in developing tools to advance democratic parliamentary reform.* PMOs have developed a variety of innovative tools that have the potential to enhance the ability of parliaments to make informed decisions and to strengthen their relationships with citizens. However, where information in open and structured formats is unavailable, PMO technologists spend much of their time converting parliamentary information into open formats by, for example, "scraping" parliamentary websites or PDF documents for data, which are then put into a structured database for analysis. When information is provided in open, structured formats, PMOs can instead focus on developing tools that add value to parliamentary information and spur citizen involvement. In addition, the international community can continue to support the efforts of PMOs and parliaments to innovate. Although some PMOs use individual member data to study broader trends in parliamentary behavior or functioning,[19] many are not yet at this point. Engaging PMOs in discussions on standards for democratic parliaments would help center attention on uses of information that would contribute to meaningful reform.

Notes

1. For more information about OGP, see the organization's website at http://www.opengovpartnership.org/about.
2. See http://visualisiert.net/parteiengesetz/index.en.html for an example of such a graphic.
3. For a timeline of the standards development process, see von Trapp (2010).
4. The conference was organized with support from the National Endowment for Democracy, the Open Society Institute, the Omidyar Network, the World Bank, and the Embassy of Mexico. More information can be found at http://www.openingparliament.org.
5. The NDI–World Bank report (Mandelbaum 2011) noted that difficulty gaining access to desired parliamentary information was the most frequently cited challenge by PMOs, which was noted by more than 63 percent of surveyed groups.

6. This statement is from the executive summary of the declaration. For the full text of the declaration, see http://www.openingparliament.org/declaration.

7. This information is contained in the "Declaration on Parliamentary Openness: Provision Commentary," a living document available at http://openingparliament.s3 .amazonaws.com/docs/declaration/commentary-20120914.pdf.

8. To view a full, updated list of supporting organizations, see http://www .openingparliament.org/organizations.

9. See the full text of the declaration at http://www.openingparliament.org /declaration.

10. For an op-ed in support of the declaration by the president of OSCE PA, see Krivokapic (2013).

11. The CPA benchmarks note that the election of parliamentary officers may take place by secret ballot and that exceptions to public committee hearings must "be clearly defined and provided for in the rules of procedure" (benchmark 3.1.4).

12. XML is a markup language that defines a set of rules for encoding documents into human-readable and machine-readable formats.

13. A URL is a specific character string, or web address, referencing a particular resource.

14. The quotation is cited in the September 2012 version of the "Declaration on Parliamentary Openness: Provision Commentary," a living document available at http:// openingparliament.s3.amazonaws.com/docs/declaration/commentary-20120914.pdf. It originally appeared in Global Centre for ICT in Parliament (2012, 103).

15. For instance, the Dutch Senate is among a number of parliaments that have reaped overall cost savings from the adoption of new technologies. See Global Centre for ICT in Parliament (2010, 78).

16. For more on Akoma Ntoso, see http://www.akomantoso.org.

17. For more on Bungeni, see http://www.bungeni.org.

18. For instance, mySociety in the United Kingdom and Fundación Ciudadano Inteligente in Chile are collaboratively developing a series of website components that any PMO can use to store and share profiles and floor speeches of members of parliament, as well as other information. See Steinberg (2012).

19. VoteWatch, which monitors the European Parliament, is an example of a website that uses individual MP voting data to illustrate trends in broader voting patterns of the parliament. See http://www.votewatch.eu.

References

Afrobarometer. 2010. *Summary of Results: Round 4.5 Afrobarometer Survey in Uganda.* Wilsken Agencies and Michigan State University, East Lansing. http://www .afrobarometer.org/files/documents/summary_results/uga_r4-5_SOR.pdf.

Banerjee, Abhijit V., Selvan Kumar, Rohini Pande, and Felix Su. 2011. "Do Informed Voters Make Better Choices? Experimental Evidence from Urban India." Working Paper, Abdul Latif Jameel Poverty Action Lab, Massachusetts Institute of Technology, Cambridge. http://www.hks.harvard.edu/fs/rpande/papers/DoInformedVoters_Nov11 .pdf.

Bruce, Tom, Eric Mill, Daniel Schuman, Josh Tauberer, and John Wonderlich. 2012. "On Public Access to Legislative Information: Recommendations to the Bulk Data

Task Force." Sunlight Foundation, Washington, DC, August 24. http://assets.sunlight-foundation.com.s3.amazonaws.com/policy/papers/THOMAS/THOMAS%20 Recommendations%20FINAL%202012-08-24.pdf.

COPA (Parliamentary Confederation of the Americas) Committee on Democracy and Peace. 2011. "The Contribution of Parliaments to Democracy: Benchmarks for the Parliaments of the Americas." Québec Secretariat of COPA, National Assembly of Québec. http://www.copa.qc.ca/eng/assembly/2011/documents/DOC-CDP-criteres -a-VF.pdf.

CPA (Commonwealth Parliamentary Association). 2006. "Recommended Benchmarks for Democratic Legislatures." CPA, London. http://wbi.worldbank.org/wbi/Data/wbi /wbicms/files/drupal-acquia/wbi/Recommended%20Benchmarks%20for%20 Democratic%20Legislatures.pdf.

———. 2011. "Recommended Benchmarks for the CPA Caribbean, Americas, and Atlantic Region Democratic Legislatures." CPA, London. http://www.cpa-caaregion .org/media/get_media.php?mediaid=caa4fafb-a31.

———. 2013. "Benchmarking 2.0: Improving Parliamentary Performance in a Tech-Enabled World." CPA, London. http://www.cpahq.org/cpahq/cpadocs/Benchmarking %202%200%20-%20Improving%20Parliamentary%20Performance%20for%20a%20 Tech-Enabled%20World.pdf.

Glader, Paul. 2012. "Internet Watchdogs: Parliament Watch Gives Voters Access to Politicians." *Spiegel Online*, May 25. http://www.spiegel.de/international/germany /german-website-lets-voters-directly-question-politicians-a-834964.html.

Global Centre for ICT in Parliament. 2010. *World e-Parliament Report 2010*. New York: United Nations.

———. 2012. *World e-Parliament Report 2012*. New York: United Nations.

Humphreys, Macartan, and Jeremy M. Weinstein. 2012. "Policing Politicians: Citizen Empowerment and Political Accountability in Uganda." Columbia University, New York. http://cu-csds.org/wp-content/uploads/2009/10/ABCDE-paper.pdf.

IPU (Inter-Parliamentary Union). 2008. "Evaluating Parliament: A Self-Assessment Toolkit for Parliaments." IPU, Geneva. http://www.ipu.org/pdf/publications/self-e.pdf.

———. 2009. "Guidelines for Parliamentary Websites." IPU, Geneva. http://www.ipu.org /PDF/publications/web-e.pdf.

Krivokapic, Ranko. 2013. "OSCE PA: Time for Parliaments to Commit to Openness." OpeningParliament.org, August 5. http://blog.openingparliament.org/post/574138 43963/osce-pa-time-for-parliaments-to-commit-to-openness.

Lee, Tom. 2012. "Scout Is Already Delivering Results." Sunlight Foundation, Washington, DC. http://sunlightfoundation.com/blog/2012/05/25/scout-is-already-delivering -results/.

Mandelbaum, Andrew G. 2011. *Strengthening Parliamentary Accountability, Citizen Engagement and Access to Information: A Global Survey of Parliamentary Monitoring Organizations*. Washington, DC: National Democratic Institute of International Affairs and World Bank. http://www.ndi.org/global-survey-parliamentary-monitoring -organizations.

Massó, Melissa Ortiz. 2013. "The Senate of Mexico Endorses the Declaration on Parliamentary Openness." OpeningParliament.org, April 30. http://blog.opening parliament.org/post/49319125315/the-senate-of-mexico-endorses-the -declaration-on.

McKenzie, Jessica. 2013. "Declaration on Parliamentary Openness Gains Wide Endorsement in Europe." TechPresident, August 7. http://techpresident.com/news/wegov/24247 /declaration-parliamentary-openness-roadmap-transparent-govt-gains-support -international.

Michener, Greg. 2012a. "Brazil's Open-Government Shock Treatment." *TechPresident*, June 27. http://techpresident.com/news/wegov/22476/brazils-open-government -shock-treatment?page=0,1.

———. 2012b. "Parliamentary Power to the People: Analyzing Online and Offline Strategies in Latin America." Latin America and Information Program, Open Society Foundations, New York, March 8. http://www.opensocietyfoundations.org/sites /default/files/parliamentary-power-20120308.pdf.

Mráček, Jakub. 2012. "Aleluja! Data Poslanecké Sněmovny k Dispozici." OKFN ČeskoPrague, October 7. http://cz.okfn.org/2012/10/07/aleluja-data-poslanecke -snemovny-k-dispozici/.

NDI (National Democratic Institute of International Affairs). 2007. "Toward the Development of International Standards for Democratic Legislatures: A Discussion Document for Review by Interested Legislatures, Donors, and International Organizations." NDI, Washington, DC. http://www.ndi.org/files/2113_gov_standards _010107.pdf.

PILDAT (Pakistan Institute of Legislative Development and Transparency). 2009. "State of Democracy in Pakistan: Evaluation of Parliament, 2008–2009. PILDAT, Islamabad.

Power, Greg, and Rebecca A. Shoot. 2012. *Global Parliamentary Report: The Changing Nature of Parliamentary Representation*. Geneva and New York: Inter-Parliamentary Union and United Nations Development Programme. http://www.ipu.org/pdf/publications /gpr2012-full-e.pdf.

SADC PF (Southern African Development Community Parliamentary Forum). 2010. "Benchmarks for Democratic Legislatures in Southern Africa." SADC PF, Windhoek. http://www.agora-parl.org/node/2777.

Steinberg, Tom. 2012. "Succeeding Means Letting Go: A Response to David Eaves." *TechPresident*, July 26. http://techpresident.com/news/wegov/22634/succeeding -means-letting-go-response-david-eaves.

Swislow, Dan E. 2012. "How PMOs Are Using the Declaration." OpeningParliament.org. http://blog.openingparliament.org/post/35844561421/how-pmos-are-using -the-declaration-part-1.

von Trapp, Lisa. 2010. "Benchmarks and Self-Assessment Frameworks for Democratic Parliaments." United Nations Development Programme, Brussels.

WBI (World Bank Institute) and UNDP (United Nations Development Programme). 2010. "Participants' Statement." International Conference on Benchmarking and Self-Assessment for Democratic Parliaments, Paris, March 2–4. http://www.ndi.org/files /Benchmarks_Conference_Participant_Statement_March2010.pdf.

Williamson, Andy. 2013. "Social Media Guidelines for Parliaments." Inter-Parliamentary Union, Geneva. http://www.ipu.org/PDF/publications/SMG2013EN.pdf.

CHAPTER 9

The African Parliamentary Index

Rasheed Draman

Introduction

The Parliamentary Centre's African Parliamentary Index (API) is an important tool for assessing parliamentary effectiveness. The API measures parliaments' level of engagement in the budget process and performance on budget oversight in selected African countries. The index adopts a narrow focus for two reasons. First, the budget process is a key area of activity for parliaments. Second, it relates closely to poverty reduction: governments allocate scarce resources that affect the lives of citizens, and in democratic states, these citizens are represented by parliamentarians. Indeed, parliamentary effectiveness requires that parliaments perform their role in the budget process and, at the same time, increase their understanding of the salient elements that are directly relevant to poverty reduction. Recognizing that parliaments' organization, powers, and effectiveness vary widely across countries, the Parliamentary Centre developed a set of indicators that measure performance of specific issues and can be aggregated into an index describing different parliaments.[1]

This chapter aims to introduce the API, and is organized as follows: The first section discusses the evolution and current state of parliaments in Africa. The second section explains the API's purpose and scope. The third section discusses the methodology used to develop the API. The final section concludes.

Parliamentary Capacities in Africa

Parliaments constitute one of the central institutions of democracy, because of their critical role in terms of legislation, oversight, and representation. As representatives of citizens' concerns and interests, parliaments oversee the executive and hold it accountable by reviewing public funds and how they are used (Africa All Party Parliamentary Group 2008, 17).

In the 1980s, numerous African countries established parliaments in conjunction with the organization of free, fair, democratically elected governments premised on multiparty democracy. Following the demise of one-party dictatorships, these parliaments began to deliberate policy, pass legislation, and strengthen links between government and the people. Many African parliaments have, albeit slowly, begun to exert the new constitutional powers that have come with the transition away from dictatorships to multiparty politics. Without doubt, African legislatures wield more power currently than at any time since independence (Barkan, Adamolekun, and Zhou 2004, 211).

However, African parliaments still face acute challenges. Many lack formal powers and clear procedures. Many also have deficient incentive structures to encourage members of parliament (MPs) and parliamentary officers to exercise their responsibilities. Indeed, according to the United Nations Economic Commission for Africa, "In terms of enacting laws, debating national issues, checking the activities of the government, and in general promoting the welfare of the people, these duties and obligations are rarely performed with efficiency and effectiveness in many African parliaments" (UNECA 2005, 127). Parliamentary strengthening activities are thus needed to develop necessary infrastructure and equipment, as well as to build capacity of the parliamentary staff, MPs, and committees.

Of course, African parliaments (and more specifically, the Parliamentary Centre's network of parliaments) encompass a range of different characteristics and mandates on budget oversight. For instance, at one end of the spectrum, some parliaments feature dominant party control and dominance by the executive, low levels of legislative activities, minimal influence on government, and little effectiveness in representing citizen concerns. At the other end of the spectrum, a growing number of parliaments have opposition groups, increased political space for debates, increased legislative activities, growing influence over government, and an increased interest and effectiveness in representing citizens.

Two factors can partly explain the variation across African parliaments. First, a parliament's character and nature is determined in part by the type of constitution upon which it is based. As noted in chapter 3, in the Westminster system in Commonwealth countries, the executive is chosen through parliamentary elections and sits in the legislature. In this system, government accountability centers on the relationship between the government and opposition parties in parliament, with MPs and parliamentary committees typically controlled through party discipline. In contrast, in the congressional system, the executive and legislative branches of government are both elected directly, and the executive sits outside parliament. In this system, accountability takes place through checks and balances between the executive and the legislature, and parliamentarians and parliamentary committees have considerable power. Finally, there are mixed systems featuring a combination of the Westminster and congressional systems.

A second factor that significantly affects a parliament's nature and operations is the type of electoral system used in the country. Constituency-based systems tend to yield majority governments, but this feature often comes at the cost of a divergence between party representation in parliament and shares of the popular vote in elections. Members concentrate their time and effort on providing services to their constituents. Proportional representation systems often have coalition governments, with continuing negotiations between parties on the makeup of the coalition. In such cases, party representation in parliament corresponds closely to voter preferences, but the connection between representatives and their constituents tends to be weaker. Although some countries have pure constituency based on proportional representation systems, others have developed mixed systems.

Purpose and Scope of the API

The API's purpose is to provide a standardized system for assessing the performance of parliaments in Africa and, potentially, elsewhere. The API has the following main objectives:

- To assess parliaments against international best practice for budget oversight
- To develop a standard and simplified system for assessing the performance of parliaments on budget oversight
- To identify priorities and tools for strengthening parliaments

The API covers two main issues: parliament's overall functions and roles and parliament's particular role in the budget process and oversight.

Under the overall functions and roles, a range of issues is covered. In particular, the API covers representation and legislation in relation to the oversight role of parliament. The index also includes data on the conditions and environment within which the institution of parliament functions. This environment includes, but is not limited to, the institutional setup that supports parliament (for example, parliament's degree of financial autonomy, the capacity of parliamentary support staff members, and the existence of a parliamentary board to give strategic direction). Lastly, the API includes issues such as gender, corruption, and the environment as cross-cutting themes as well as independent issues in their own right. The API considers both ex ante and ex post parliamentary involvement in the budget process by assessing parliament's influence in budget formulation and ability to scrutinize past expenditures.

Approach and Methodology

Organization

To gather data for the API, independent country assessors (ICAs) oversaw the completion of self-assessments by a sample of experienced parliamentarians and parliamentary staff members. To achieve a representative sample, ICAs targeted MPs from opposition and governing political parties, as well as parliamentary staff members, and took gender balance into consideration.

For the first round of self-assessments, ICAs presented the API's concept and scope and, when necessary, assisted participants in finding a common understanding of the exercise's purpose and each group member's role. ICAs introduced the primary toolkit and described each indicator of the API in detail (see annex 9A for a copy of the toolkit). Participants then discussed possible scores for each indicator, evidence to support their ratings, and recommendations for capacity enhancement.

Self-assessment participants also contributed to the construction of a weighting table by ranking pairs of indicators. These rankings were used to tabulate an average weight for each indicator. The weighted-capacity average was calculated by multiplying the calculated priority weight by the average score over the maximum possible score (which was four).

As a validation measure, the API gathered assessments from relevant nongovernmental organizations (NGOs) and civil society organizations (CSOs) about the work of parliamentarians. The aim of this validation exercise was twofold. The first objective was to generate complementary insights and credible feedback on parliamentary capacity from informed representatives of CSOs. The second motivation was to familiarize other stakeholders with best practices in budget oversight and to share information on the constraints facing parliaments.

Participation

The API self-assessment exercise was undertaken in five countries in Sub-Saharan Africa (Benin, Ghana, Kenya, Tanzania, and Uganda). Participants in the self-assessment workshops and validation exercises were as follows:

- In Benin, 26 participants (8 MPs and 18 parliamentary staff members) took part in the self-assessment. Seven independent observers also provided support during the working group session. Four working groups were formed to facilitate the assessment exercise. Sixteen civil society participants took part in the validation exercise (with the support of six independent observers).

- In Ghana, 33 people participated in the parliamentary assessment exercise (16 MPs and 17 parliamentary staff members). In addition, 20 participants from civil society undertook the exercise to validate the conclusions reached by parliamentarians. Validation participants were primarily representatives from key NGOs, think tanks, and academic and research institutions that engage parliament regularly.

- In Kenya, 23 participants (10 MPs and 13 parliamentary staff members) took part in the parliamentary self-assessment. Five staff members from the Parliamentary Centre, an independent assessor, and the independent assessor's assistant offered support throughout the working group session. Also, 13 representatives from NGOs and CSOs participated in the civil society validation workshop, supported by an independent assessor and an assistant.

- In Tanzania, the parliamentary self-assessment exercise attracted 19 participants (11 MPs and 8 parliamentary staff members). An independent assessor and an assistant facilitated the workshop. Eight representatives from NGOs and CSOs validated the conclusions reached by parliamentarians through the follow-up exercise.

- In Uganda, about 24 MPs and parliamentary staff members undertook the parliamentary self-assessment exercise. An independent assessor and an assistant supported this working group. The civil society validation exercise then involved seven CSOs with a history of association with the Parliament of Uganda.

- In all country assessments, Parliamentary Centre representatives provided technical and logistical support.

Areas Assessed

The self-assessment tool used by the MPs and parliamentary staff members covers five core areas: representation, legislation, oversight functions, institutional capacity, and institutional integrity.

Self-assessment questions are largely quantitative, requiring respondents to make judgments and score each variable or indicator on a four-point scale. On this scale, four denotes a high level of parliament capacity, three shows a moderate level of capacity, two indicates the existence of a basic level of capacity, and one signals a clear need for capacity development (see annex 9A).

This quantitative approach makes it is possible to undertake a comparative analysis of different country experiences, to highlight good practice and lessons learned, and to make specific recommendations for improving parliament's role in the budget process.

Conclusion

In summary, the Parliamentary Centre developed the API to provide a standardized system for assessing the performance of African parliaments. This set of indicators measures parliaments' level of engagement in the budget process and effectiveness of budget oversight in selected African countries.

Although the API initiative is not the first of its kind in Africa, its added value lies in the fact that parliaments themselves were the key drivers of the assessment process. The high level of interest demonstrated by the participating parliaments gives cause for optimism with regard to the sustainability of the API process.

The Parliamentary Centre seeks to contribute to effective participatory democracy in Africa, with the continued sponsorship of the Canadian International Development Agency under the African Parliamentary Strengthening Program and the significant support and participation of all the partner parliaments and their staffs in its programs.

Benchmarking and Self-Assessment for Parliaments • http://dx.doi.org/10.1596/978-1-4648-0327-7

Annex 9A: Self-Assessment for African Parliaments—Parliaments' Role in the Budget Process

Legislature: Date:

Honorable members of the legislature, this self-assessment tool examines the level of engagement of your legislature with the budget process and the oversight of government expenditures in your country. The African Parliamentary Strengthening Program (APSP) is a five-year capacity-strengthening program that supports seven partner parliaments in developing and implementing strategies that strengthen their overall role and engagement in the national budget process. This assessment is part of a larger framework to monitor results of this program, and it aims to provide stakeholders with a simplified and standard assessment of partner parliaments' role in the budget process. It is also intended to help identify gaps that would inform programming under the APSP.

The tool focuses on assessing the legislature as a whole (not individuals) on core capacity elements in effective budgeting and oversight. An independent assessor should facilitate the self-assessment. For each of the identified capacity areas, please rate your legislature on a scale of one to four, and then continue with the qualitative information to explain your rating. If a capacity element does not apply to your legislature, select "N/A." In the evidence column, provide as much evidence as possible to support your rating and provide a reference for your response. In the recommendation column, the assessment team should suggest ways to address an identified capacity gap.

Assessors: Please list the names of MPs and staff members engaged in this assessment.

Legislators	Parliamentary staff members	Independent observers and assessors

Representation	Level 4: High level of capacity in place	Level 3: Moderate level of capacity in place	Level 2: Basic level of capacity in place	Level 1: Clear need for increased capacity	Rating	Evidence	Recommendation
Accessibility							
1. Openness of the legislature to citizens and the media	The legislature is accessible to citizens and the media. This accessibility is guided by a framework and the legislature's communication strategy.	The legislature is open to citizens and the media. A communication strategy exists but does not focus on accessibility by citizens and the media.	The legislature is only partially open to citizens and the media. Accessibility is usually in response to pressure from organized groups. No communication strategy or framework exists to structure and guide accessibility.	The legislature is not open to citizens and the media. No communication strategy exists in this area.			
2. Use of a nonpartisan media relations facility by the legislature	The legislature has a nonpartisan media center. This media center gives access to all media houses, is perceived to be nonpartisan, and is guided by a code of conduct.	The legislature has a media center that is supposed to be nonpartisan. This media center gives access to all media houses but is perceived to be partisan. The legislature is not guided by a code of conduct that gives access to all media houses.	The legislature does not have a nonpartisan media center. A media relations unit exists, but the unit has no policy to guide the legislature's interaction with the media and the public.	The legislature does not have a nonpartisan media center. No unit exists for media relations.			
3. Mechanisms to promote public understanding of the legislature's work	A carefully structured process exists and is followed to promote the public's understanding of the legislature's work.	Mechanisms exist to promote public understanding of the legislature's work. These mechanisms are not followed and not well structured.	No mechanisms exist to promote the public's understanding of the legislature's work. However, the legislature makes an attempt to promote public understanding of its work sometimes.	No mechanisms exist to promote the public's understanding of the legislature's work. Very little attempt is made to promote this interest among the public.			

table continues next page

Representation	Level 4: High level of capacity in place	Level 3: Moderate level of capacity in place	Level 2: Basic level of capacity in place	Level 1: Clear need for increased capacity	Rating	Evidence	Recommendation
4. Timely provision of information to the public on the budget	Information is provided to the public in a timely manner regarding budgets under consideration by the legislature.	Information on budgets under consideration by the legislature is provided to the public but not in a timely manner.	Information on the budget is provided to the public as and when the legislature deems it necessary.	Information on the budget is not provided to the public.			
5. Promoting citizens' knowledge and understanding of the role of members of parliament (MPs) in the budget process	Carefully structured processes exist and are followed to promote citizens' knowledge and understanding of MPs' role in the budget process.	Mechanisms exist to promote citizens' knowledge and understanding of MPs' role in the budget process. However, these mechanisms are not well structured and not followed.	No mechanisms exist to promote citizens' knowledge and understanding of MPs' role in the budget process. However, the legislature sometimes makes an attempt to promote citizens' understanding of the role of MPs in the budget process.	No mechanisms exist to promote citizens' knowledge and understanding of MPs' role in the budget process. Very little attempt is made to promote this interest among the public.			
6. Relationship between (a) parliament and (b) civil society organizations (CSOs) and other related institutions	There are clear guidelines in the rules of procedure and other laws governing the relationship between the legislature and CSOs and other institutions. The guidelines provide entry points for CSOs' input into the legislature's work.	There are clear guidelines in the rules of procedure and other laws governing the relationship between the legislature and CSOs and other institutions. However, these guidelines do not provide entry points for CSOs' input into the legislature's work.	There are no clear guidelines in the rules of procedure and other laws governing the relationship between the legislature and CSOs and other institutions. The relationship is ad hoc and determined by the legislature.	There are no guidelines in the rules of procedure and other laws governing the relationship between the legislature and CSOs and other institutions. CSOs have no opportunity to provide input into the legislature's work.			

table continues next page

Legislative function

Legal mandate

Representation	Level 4: High level of capacity in place	Level 3: Moderate level of capacity in place	Level 2: Basic level of capacity in place	Level 1: Clear need for increased capacity	Rating	Evidence	Recommendation
7. Lawmaking, including the appropriations act	The power of the legislature to make laws and acts, including the appropriations act, is contained in the constitution.	The power of the legislature to make laws and acts, including the appropriations act, is contained in an act.	The power of the legislature to make laws and acts, including the appropriations act, is based on convention.	The power of the legislature to make laws and acts, including the appropriations act, has no legal backing.			
8. Power to amend the appropriations bill	The legislature has unlimited power to amend the appropriations bill.	The legislature has power to amend the appropriations bill but cannot exceed the budget ceiling.	The legislature can only amend the appropriations bill with the consent of the minister for finance or the executive.	The legislature does not have power to amend the appropriations bill.			
9. Opportunities for public input into the legislative process	Adequate opportunities exist for citizens to provide input into any legislative process. These opportunities are contained in the rules of procedure or other laws or instruments and are made public.	Adequate opportunities exist for citizens to provide input into any legislative process and are made public. However, these opportunities are not backed by legislation.	Opportunities exist for citizens to provide input into any legislative process. However, these opportunities are not backed by legislation and are not made public.	There are no opportunities for citizens to provide input into any legislative process.			
10. Mechanisms to track legislation	Adequate mechanisms exist for the legislature to track legislation that has been enacted. The legislature has access to resources to provide evidence on the effect of specific legislation.	Mechanisms exist for the legislature to track legislation that has been enacted. Some resources exist to provide evidence on the effect of specific legislation, but they are inadequate.	Some mechanisms exist for the legislature to track legislation that has been enacted, but they are inadequate and need to be reviewed. Resources that provide evidence on the effect of legislation are lacking.	No mechanisms exist for the legislature to track legislation that has been enacted.			

table continues next page

Financial function

The budget review and hearing

Representation	Level 4: High level of capacity in place	Level 3: Moderate level of capacity in place	Level 2: Basic level of capacity in place	Level 1: Clear need for increased capacity	Rating	Evidence	Recommendation
11. Period for the budget review by the legislature	The legislature has at least 3 months to review the budget.	The legislature has not less than 2 months but not more than 3 months to review the budget.	The legislature has not less than 1 month but not more than 2 months to review the budget.	The legislature has 1 month or less to review the budget.			
12. Existence of an appropriations or budget committee	There is an appropriations or budget committee whose sole mandate is to review the budget.	There is an appropriations or budget committee, but it shares the mandate of the budget review with other standing committees.	There is no appropriations or budget committee. A special or ad hoc committee reviews the budget.	There is no appropriations or budget committee. The budget review is done at plenary of the legislature.			
13. Public hearings on the budget	The appropriations committee and other committees hold public hearings on the budget during which evidence from the executive and the public is taken.	The appropriations committee and other committees hold public hearings on the budget, but the hearings are a one-way presentation by the executive. The public has no input.	The appropriations committee and other committees hold public hearings only when the chair and members so decide.	The appropriations committee and other committees are not permitted to hold public hearings on the budget.			
14. Process for citizens' participation in the budget process	The process for citizens' participation in the budget process is effective. The process is well documented, is an integral part of the legislature's communication strategy, and is known to the public.	A process for citizens' participation in the budget process exists. The process is well documented, but it is not publicized and is therefore not known to the public.	Some processes for the participation of citizens in the budget process exist, but they are not documented and are not known to the public.	No process exists for citizens to participate in the budget process.			

table continues next page

Representation	Level 4: High level of capacity in place	Level 3: Moderate level of capacity in place	Level 2: Basic level of capacity in place	Level 1: Clear need for increased capacity	Rating	Evidence	Recommendation
15. Authority to amend budget presented by the executive	The legislature has authority in law to amend the budget presented by the executive, including spending and revenue proposals.	The legislature has the authority to make proposals for amendment. These proposals must, however, be backed by funding sources in case of an upward review.	The legislature does not have the authority in law to make amendments to the budget but may sometimes negotiate with the executive for amendments to be made.	The legislature cannot and does not make amendments to the budget presented by the executive.			
16. Power to send back proposed budget for review	The rules of procedure or other laws empower the legislature to send the budget back to the executive for review. This provision is often exercised.	The rules of procedure or other laws empower the legislature to send the budget back to the executive for review. This provision is usually not exercised.	There are no rules or laws that empower the legislature to send the budget back to the executive for review. However, there are informal arrangements for some aspect to be sent to the executive for review. Such a review is seldom done.	The legislature does not have the power to send the budget back to the executive for review.			
17. Amendments on spending and revenue proposals	Amendments made by the legislature on spending and revenue proposals are binding on the executive.	Amendments made by the legislature on spending and revenue proposals are binding on the executive, but the executive usually finds an excuse not to implement the amendments.	Amendments made by the legislature on spending and revenue proposals are not binding on the executive, but the executive usually implements the amendments.	Amendments made by the legislature on spending and revenue proposals are not binding on the executive, and the executive does not implement these amendments.			
18. Information in the appropriations (expenditure budget) approved by the legislature	The appropriations approved by the legislature contain detailed information	The appropriations approved by the legislature contain information on all ministries,	The appropriations approved by the legislature contain information on selected ministries, departments,	The appropriations approved by the legislature do not contain relevant information.			

table continues next page

Representation	Level 4: High level of capacity in place	Level 3: Moderate level of capacity in place	Level 2: Basic level of capacity in place	Level 1: Clear need for increased capacity	Rating	Evidence	Recommendation
	on all ministries, departments, and agencies.	departments, and agencies, but that information is not detailed.	and agencies, but that information is not detailed.				
Budget act and budget office							
19. Existence of a budget act	There is a budget act that clearly defines a role for the legislature in the budget process.	There is a budget act, but it needs revision to make it more relevant to the needs of modern times. Processes have been initiated to review the act.	There is no budget act, but the legislature follows best practices and plays its expected role in the budget process. A process to enact a budget act has started.	There is no budget act, and the legislature does not play a major role in the budget process. There is no process in place to enact a budget act.			
20. Existence of a budget office	The legislature has a budget office that is established by law (budget act).	The legislature has a budget office, but it has no legal backing.	There is no budget office, but a unit exists that provides research support on the budget to the legislature.	There is no budget office or research unit available to the legislature.			
21. Resourcing of the budget office	The legislature's budget office has qualified, competent officers and is equipped to efficiently and effectively deliver to the legislature.	The legislature's budget office has competent officers, but they lack the necessary resources to function effectively.	The legislature's budget office is well equipped but lacks competent officers.	The budget office lacks the necessary personnel and equipment to perform its job.			
22. Access to information from central government departments and the private sector	The budget office has power to call for information and documents from government departments and the private sector and in good time (power of subpoena).	Although the budget office has power of subpoena, this power is sometimes not respected by government departments and the private sector.	The budget office has no legal backing for requesting information from central government agencies and the private sector. The agencies respond to the budget office's requests at their convenience.	The budget office has no legal backing for requesting information from central government agencies and the private sector. The budget office's requests are not respected.			

table continues next page

Representation	Level 4: High level of capacity in place	Level 3: Moderate level of capacity in place	Level 2: Basic level of capacity in place	Level 1: Clear need for increased capacity	Rating	Evidence	Recommendation
23. Legislature's consideration of estimates for defense and intelligence services	The legislature (or the appropriate committee) considers and approves the budget estimates for defense and intelligence services and is given full disclosure on these estimates.	The legislature (or the appropriate committee) considers and approves the budget estimates for defense and intelligence services, but there is no full disclosure on the estimates.	A special committee considers and approves the estimates for defense and intelligence services. The committee's report is not discussed at the plenary of the legislature.	The legislature does not consider or approve the budget for defense and intelligence services.			
Periodic review of the budget							
24. Budget reviews	The budget is reviewed every year by the executive (number of reviews in a year and types).	The budget is reviewed by the executive but not every year (number of reviews and types).	The budget is seldom reviewed by the executive (number of reviews and types).	The budget is never reviewed.			
25. Legislative approval of reviews	All budget reviews are presented to and approved by the legislature.	All budget reviews are presented to the legislature but do not require its approval.	Budget reviews are presented to the legislature as and when the executive deems necessary.	Budget reviews are not presented to the legislature and do not require its approval.			
26. Time allocated for approval of reviewed budget	Adequate time is allocated for the consideration of the reviewed budget both at plenary and at committees.	Adequate time is allocated for the consideration of the reviewed budget but only at plenary.	Only limited time is allocated for the consideration of the reviewed budget.	The legislature has no scheduled time to consider the reviewed budget.			
Oversight function							
Oversight committees							
27. Existence of oversight committees	The legislature's oversight function is performed by all sector-related committees and other special committees.	The legislature's oversight function is performed by special committees.	The legislature's oversight function is performed by only one specialized committee.	The legislature's oversight function is performed by the legislature at plenary.			

table continues next page

Representation	Level 4: High level of capacity in place	Level 3: Moderate level of capacity in place	Level 2: Basic level of capacity in place	Level 1: Clear need for increased capacity	Rating	Evidence	Recommendation
28. Investigative powers of oversight committees	Oversight committees have investigative powers over budgetary issues and government spending. These powers are enshrined in the rules of procedure or other laws and are regularly enforced.	Oversight committees have investigative powers over budgetary issues and government spending. These powers are enshrined in the rules of procedure or other laws but are not regularly enforced.	Oversight committees have investigative powers over budgetary issues and government spending, but these powers are not backed by law.	Oversight committees do not have investigative powers over budgetary issues and government spending.			
29. Oversight of spending by state enterprises	Oversight committees sufficiently oversee the expenditures of state-owned enterprises. The committees can call for special audits or invite officers of respective state-owned enterprises to testify before them.	Oversight committees sufficiently oversee the expenditures of state-owned enterprises. The committees can invite officers of respective state-owned enterprises to testify before them but cannot at any point in time call for special audits.	Oversight committees do not sufficiently oversee the expenditures of state-owned enterprises. The committees cannot call for special audits nor invite officers of respective state-owned enterprises to testify before them.	The legislature does not oversee the expenditures of state-owned enterprises.			
30. Mechanisms for oversight committees to obtain information from the executive	Sufficient mechanisms exist for committees to obtain information from the executive to exercise their oversight function in a meaningful way. These mechanisms have proven time and again to work well.	Mechanisms exist for committees to obtain information from the executive to exercise their oversight function, but these mechanisms are not efficient.	Mechanisms do not exist for committees to obtain information from the executive to exercise their oversight function. The legislature recognizes this gap and is taking steps to address it.	There are no mechanisms for committees to obtain information from the executive to exercise their oversight function in a meaningful way.			

table continues next page

Representation	Level 4: High level of capacity in place	Level 3: Moderate level of capacity in place	Level 2: Basic level of capacity in place	Level 1: Clear need for increased capacity	Rating	Evidence	Recommendation
31. Power of oversight committees to follow up on recommendations	Oversight committees have adequate powers in law to request and receive response on actions taken by the executive on the committees' or parliament's recommendations.	Oversight committees have adequate powers in law to request and receive response on actions taken by the executive on the committees' or parliament's recommendations but do not receive frequent updates on action taken.	Oversight committees do not have adequate powers to request and receive response on actions taken by the executive on recommendations, but sometimes they receive a report from the executive on action taken.	Oversight committees do not have power to request and receive response on actions taken by the executive on the committees' or parliament's recommendations.			
32. Access to resources by oversight committees	Oversight committees are adequately resourced to undertake their activities. The committees have separate budgets.	Oversight committees are adequately resourced but do not have separate budgets. Committees apply to the speaker and other leadership for resources and funds for their activities.	Oversight committees are not adequately resourced and do not have separate budgets. Committees apply to the speaker and other leadership for resources for their activities.	Oversight committees are poorly resourced and do not have separate budgets for their activities.			
33. Opportunities for minority and opposition parties	Oversight committees provide meaningful opportunities for minority and opposition parties to engage in effective oversight of government expenditures.	Oversight committees provide limited opportunities for minority and opposition parties to engage in effective oversight of government expenditures.	Oversight committees are dominated by the ruling party. Minority and opposition parties have very limited opportunities to engage in oversight of government expenditures.	Oversight committees are dominated by the ruling party. Minority and opposition parties have no opportunity to engage in oversight of government expenditures.			

table continues next page

Public accounts committee

Representation	Level 4: High level of capacity in place	Level 3: Moderate level of capacity in place	Level 2: Basic level of capacity in place	Level 1: Clear need for increased capacity	Rating	Evidence	Recommendation
34. Existence of a public accounts committee (PAC)	The legislature has a PAC that examines government expenditures and is established by the constitution or an act of parliament.	The legislature has a PAC that examines government expenditures and is established by the rules of procedure (standing orders).	The legislature has a PAC that examines government expenditures, but it is established by convention.	The legislature has no PAC to examine government expenditures.			
35. Chair of the PAC	The PAC is chaired by a member who does not belong to the party in government. The law or rules of procedure provide for this arrangement.	The PAC is chaired by a member who does not belong to the party in government. The law or rules of procedure do not provide for this arrangement, but it has been adopted by convention.	The chair of the PAC is elected by members of the committee and can be from the party in government or another party.	The PAC is chaired by a member from the party in government.			
36. Rights and powers of the PAC	The PAC has power to subpoena witnesses and documents, and this power is backed by law.	The PAC can subpoena witnesses and documents. This power is not backed by a specific law but has been adopted by convention.	The PAC can invite witnesses and request documents, but it cannot compel compliance.	The PAC cannot invite witnesses or request documents.			
37. Attendance by ministers	Ministers are mandated to attend the PAC's meetings.	Ministers may attend the PAC's meetings, but attendance is not mandatory.	Ministers are not permitted to attend the PAC's meetings. Only public and civil servants are required to attend such meetings.	Neither ministers nor civil servants are mandated to attend the PAC's meetings.			
38. Openness of the PAC proceedings	The PAC is required by law to hold its proceedings in public, and the public can provide input during such proceedings.	The PAC is required by law to hold its proceedings in public, but the public cannot provide input during such proceedings.	The PAC may hold its proceedings in public if the chair and members so decide, but the public cannot provide input during such proceedings.	The PAC's proceedings are held in camera (not open to the public).			

table continues next page

Representation	Level 4: High level of capacity in place	Level 3: Moderate level of capacity in place	Level 2: Basic level of capacity in place	Level 1: Clear need for increased capacity	Rating	Evidence	Recommendation
39. Consideration of reports of the auditor general (AG)	The PAC considers all AG reports in a timely manner.	The PAC considers all AG reports but not in a timely manner.	The PAC considers some AG reports but not in a timely manner.	The PAC rarely considers the AG reports.			
40. Independent investigations	The PAC can independently investigate any matter of public interest.	The PAC can independently investigate any matter of public interest subject to the approval of the legislature.	The PAC can independently investigate any matter of public interest, but the investigation must be approved by the speaker.	The PAC cannot initiate any independent investigation.			
41. Recommendations of the PAC	The executive is bound by law to implement the PAC's recommendations, and this provision is strictly enforced.	The executive is bound by law to implement the PAC's recommendations, but this provision is not strictly enforced.	The executive is not bound by law to implement the PAC's recommendations but nonetheless implements most recommendations.	The executive is not bound by law to implement the PAC's recommendations and rarely implements such recommendations.			
42. Mechanisms for tracking the PAC's recommendations	Adequate mechanisms exist for the PAC to track the implementation of its recommendations, and such tracking can be accessed and verified by the public.	Adequate mechanisms exist for the PAC to track the implementation of its recommendations, but such tracking cannot be accessed and verified by the public.	Some mechanisms exist for the PAC to track the implementation of its recommendations, but they are rarely used.	No mechanism exists for the PAC to track the implementation of its recommendations.			
43. Resourcing of the PAC	The PAC is adequately resourced to undertake its activities. The committee has a separate budget.	The PAC is adequately resourced but has no separate budget. The committee applies to the speaker or other leadership for funds.	The PAC is not adequately resourced and has no separate budget. The committee depends on the bureaucracy (parliamentary service) for resources.	The PAC is poorly resourced and has no separate budget for its activities.			
44. Collaboration with anticorruption institutions	The PAC has a formal and strong collaboration with other anticorruption institutions.	The PAC has a good but informal collaboration with anticorruption institutions.	The PAC has an informal collaboration with a limited number of anticorruption institutions.	The PAC has no relationship with anticorruption institutions.			

table continues next page

Audit

Representation	Level 4: High level of capacity in place	Level 3: Moderate level of capacity in place	Level 2: Basic level of capacity in place	Level 1: Clear need for increased capacity	Rating	Evidence	Recommendation
45. Appointment of the AG	The AG is appointed by and responsible to the legislature.	The AG is appointed by the president and confirmed by and responsible to the legislature.	The AG is appointed by the president in consultation with a special body (such as an institute of accountants, council of state, or public service commission).	The AG is appointed by and reports to the president.			
46. Submission of AG reports	The AG submits all reports to the legislature.	The AG submits many reports to the legislature.	The AG submits a few reports to the legislature.	The AG does not submit reports to the legislature.			
47. Regularity and timeliness of AG reports	The legislature receives regular and timely AG reports.	The legislature receives regular but not timely AG reports.	The legislature receives timely but not regular AG reports.	The legislature does not receive regular and timely AG reports.			
48. Publication of AG reports	AG reports are deemed public immediately after they are issued.	AG reports are deemed public after they are laid before the legislature.	AG reports are deemed public after they have been considered by the PAC.	The AG and the legislature determine when to make such reports public.			
49. Request for audit	The legislature can request the AG to conduct special audits on its behalf, and the AG is obliged to comply.	The legislature can request the AG to conduct special audits on its behalf, but the AG is not obliged to comply.	The legislature can request the AG to conduct special audits, but the legislature must pay for such audits.	Only the president can request the AG to conduct special audits.			
50. The AG's resources and authority	The AG has adequate resources and legal authority to conduct audits in a timely manner.	The AG has limited resources but has legal authority to conduct audits in a timely manner.	The AG has adequate resources but no legal authority to conduct audits in a timely manner.	The AG does not have adequate resources and legal authority to conduct audits in a timely manner.			

Institutional capacity of the institution of parliament

Financial and material resources

Representation	Level 4: High level of capacity in place	Level 3: Moderate level of capacity in place	Level 2: Basic level of capacity in place	Level 1: Clear need for increased capacity	Rating	Evidence	Recommendation
51. Power of the legislature to determine its own budget	The legislature determines its budget for the year, and the executive cannot vary it.	The legislature determines its budget for the year, but the executive provides funds as and when funds are available.	The budget for the legislature is subject to the president's approval.	The minister for finance determines the legislature's budget.			

table continues next page

Representation	Level 4: High level of capacity in place	Level 3: Moderate level of capacity in place	Level 2: Basic level of capacity in place	Level 1: Clear need for increased capacity	Rating	Evidence	Recommendation
52. Logistics available to the legislature	The legislature has good logistics, including office space to enable it to perform its functions.	The legislature has basic logistics, including office space to enable it to perform its functions.	The legislature has basic logistics but lacks adequate office space for its functions.	The legislature lacks the basic logistics and office space to enable it to perform its functions.			
53. MPs' resources for constituency development and activities	Each MP has a constituency development fund that is used for development projects in the constituency and is independently managed by the MP.	Each MP has a constituency development fund that is used for development projects in the constituency and is managed jointly by the legislature and the MP.	Each MP has a constituency development fund that is used for development projects in the constituency and is managed by the MP and the local authority.	MPs have no constituency development fund that is used for development projects in the constituency.			
54. Mechanism for receiving and coordinating technical assistance	The legislature has a structured system for receiving technical and advisory assistance from external sources. A fully staffed donor coordination unit exists.	The legislature has a structured system for receiving technical and advisory assistance from external sources. However, there is no specific desk or unit for such purpose.	Coordination of technical assistance to the legislature is ad hoc. It is difficult to have a complete overview of technical assistance.	The legislature does not have a structured system for receiving technical and advisory assistance from external sources.			
Human resources							
55. Equal opportunity employment	The legislature does not discriminate in its recruitment of staff members on the basis of race, ethnicity, religion, gender, disability, or party affiliation.	The legislature does not discriminate in its recruitment of staff members. However, there is the perception that the ruling party strongly influences the recruitment process.	Though the legislature does not usually discriminate in its recruitment of staff members on the basis of race, ethnicity, religion, gender, disability, or party affiliation, nondiscrimination is sometimes overlooked.	The legislature recruits staff members on the basis of race, ethnicity, religion, gender, disability, and according to party affiliation. Staffing is highly polarized.			

table continues next page

Representation	Level 4: High level of capacity in place	Level 3: Moderate level of capacity in place	Level 2: Basic level of capacity in place	Level 1: Clear need for increased capacity	Rating	Evidence	Recommendation
56. Research and other support staff	The legislature has highly competent specialists, researchers, and other staff members that provide research and other information in real time, including position papers on topical issues.	The legislature has research and support staff members, but they lack the requisite background and tools to enable them to provide MPs with information in real time.	The legislature has some support staff members, but they are not specialists and meet only basic information needs of MPs.	The legislature has no research officers.			
Transparency and integrity							
57. Existence and compliance with a code of conduct	The legislature has a code of conduct that guides the MPs' behavior and actions. The code is backed by legislation and is strictly enforced.	The legislature has a code of conduct that guides the MPs' behavior and actions, but it is not backed by legislation. It is, however, enforced.	The legislature has no specific code of conduct. There are, however, some provisions in the rules of procedure that guide the MPs' conduct.	The legislature has no code of conduct nor provisions in the rules of procedure to guide the MPs' conduct.			
58. Maintenance of high standards of accountability, transparency, and responsibility	MPs maintain high standards of accountability, transparency, and responsibility in the conduct of public and parliamentary work.	MPs maintain some standards of accountability, transparency, and responsibility in the conduct of public and parliamentary work.	MPs maintain low standards of accountability, transparency, and responsibility in the conduct of public and parliamentary work.	MPs maintain very low standards of accountability, transparency, and responsibility in the conduct of public and parliamentary work.			
59. Mechanisms for anticorruption activities	Anticorruption networks exist, and MPs are free and encouraged to join. MPs are motivated to participate in anticorruption activities.	Anticorruption networks exist, and MPs are free and encouraged to join, but they are not motivated to participate in anticorruption activities.	No formal anticorruption networks exist, but MPs come together on anticorruption issues. There is little motivation for networking.	No anticorruption network exists, and MPs are not permitted to engage in such networks.			

table continues next page

Representation	Level 4: High level of capacity in place	Level 3: Moderate level of capacity in place	Level 2: Basic level of capacity in place	Level 1: Clear need for increased capacity	Rating	Evidence	Recommendation
60. Mechanisms to prevent, detect, and discipline MPs and staff members engaged in corrupt practices	Efficient and effective mechanisms exist to detect and prevent corrupt practices among MPs and staff members and to bring to justice any person engaged in such activities. These mechanisms are known to all.	Mechanisms exist to detect and prevent corrupt practices among MPs and staff members and to bring to justice any person engaged in such activities. However, these mechanisms are not efficient or effective.	There are no mechanisms to detect and prevent corrupt practices among MPs and staff members and to bring to justice any person engaged in such activities. MPs and staff members are guided by their own ethical principles.	There are no mechanisms to detect and prevent corrupt practices among MPs and staff members and to bring to justice any person engaged in such activities.			
61. Declaration of assets and business interests	MPs are required by law and the rules of procedure to declare their assets and business interests, and they strictly comply with this requirement.	MPs are required by law and the rules of procedure to declare their assets and business interests, but the provision is not enforced and only a few MPs comply.	There is no law that requires MPs to declare their assets and business interests, but there is a system that encourages MPs to do so voluntarily.	MPs are under no obligation to declare their assets and business interests, and there is no system that encourages such disclosures.			

195

Note

1. The African Parliamentary Strengthening Program for Budget Oversight is a five-year capacity-strengthening program for seven partner parliaments (in Benin, Ghana, Kenya, Senegal, Tanzania, Uganda, and Zambia). The program, funded by the Canadian International Development Agency and implemented by the Parliamentary Centre's African Program, supports the seven partner parliaments in their efforts to develop and implement strategies that strengthen their overall role and engagement in the national budget process. The Parliamentary Centre designed the API to provide a standard and simplified system for assessing parliaments' performance in Africa, particularly in the seven partner parliaments.

References

Africa All Party Parliamentary Group. 2008. "Strengthening Parliaments in Africa: Improving Support." Africa All Party Parliamentary Group, London.

Barkan, Joel D., Ladipo Adamolekun, and Yongmei Zhou. 2004. "Emerging Legislatures: Institutions of Horizontal Accountability." In *Building State Capacity in Africa: New Approaches, Emerging Lessons*, edited by Brian Levy and Sahr Kpundeh, 211–56. Washington, DC: World Bank.

UNECA(United Nations Economic Commission for Africa). 2005. *African Governance Report 2005*. Addis Ababa: UNECA.

CHAPTER 10

Assessing Parliamentary Oversight in Sri Lanka

Raja Gomez

Introduction

This chapter assesses the capabilities of the Sri Lankan parliament through the lens of two legislative evaluation frameworks: the Self-Assessment Toolkit for Parliaments (IPU 2008) of the Inter-Parliamentary Union (IPU) and the benchmarks (CPA 2006) of the Commonwealth Parliamentary Association (CPA).

These evaluation rubrics were initially applied as part of a 2008–09 study for the World Bank, which sought to assist in the assessment of the oversight operations of legislatures regarding the budget process (Gomez 2008a, 2008b, 2009). This chapter updates the information collected for the prior study in light of recent changes that have occurred in the Sri Lankan political paradigm, foremost among which is the conclusion of the civil war in 2009, which ended with the pacification of the Liberation Tigers of Tamil Eelam (LTTE).

Therefore, after providing an overview of the Sri Lankan parliament, this chapter systematically covers the salient metrics of both the IPU and the CPA assessments, highlighting Sri Lanka's strengths and weaknesses according to both evaluation tools in the most succinct manner possible given the comprehensive breadth of both assessment schemes.

Background

Sri Lanka has had a system of parliamentary government since gaining independence from the United Kingdom in 1948 under its former name of Ceylon. It has been a member of the Commonwealth of Nations since that time and a member of the United Nations since 1955. In the initial postindependence period, the country remained a British dominion, and its parliament consisted of a senate and a house of representatives. In 1972, the country became a republic under the

While taking responsibility for all material presented in the original studies, Raja Gomez acknowledges the assistance of Usman Chohan in the preparation of this chapter for publication in its current form.

name of Sri Lanka with a constitutional (that is, ceremonial rather than executive) president and a single-chamber parliament. In 1978, a new constitution was adopted that created an executive presidency and replaced the first-past-the-post election system with an extensive proportional representation mechanism that applied to all levels of public governance. The practice and procedures of the parliament of Sri Lanka have developed in their own way since independence, but their derivation from those of the United Kingdom's House of Commons can still be seen. Parliament meets every other week (with a few predetermined exceptions) throughout the year, with no vacations or recesses; only a prorogation or dissolution interrupts its activities.

Much of parliament's energy and, indeed, that of the country as a whole have been absorbed by the war with the LTTE rebel group, which finally ended with its military defeat in 2009. This context should be borne in mind when interpreting the work of parliament and the ways in which the public and the media perceive parliament. Most important, this history has deprioritized the debate on what would otherwise be key fields, including finance and taxation. It forms, in other words, an important backdrop to the discussion of the relevance and applicability of the frameworks discussed in this chapter.

IPU Toolkit for Self-Assessment

The IPU self-assessment was carried out during a week in February–March 2009, and some further clarification was obtained in May of that year. The findings have been updated and some additional material included in light of recent developments and further experience in using the toolkit. (For further information and context on the IPU toolkit, see chapter 2 of this volume.)

This section is based on the views of the senior staff of the parliament, and its purpose is to describe how the self-assessment exercise was conducted, the lessons learned in the process, and the type of outcome that resulted. No mandatory procedures are specified for use of the toolkit: the IPU itself states that "each parliament will decide for itself how to approach the self-assessment exercise" (IPU 2008, 8). However, the guidance material issued by the IPU includes various helpful suggestions regarding the possible makeup of a participant group, the role of a facilitator, the timeframe of work, documentation, and possible outcomes. Use of an essentially numerical rating system was a novel experiment at the time the exercise was conducted.

Twelve individual members of the senior staff participated in the self-assessment, including the deputy secretary general of parliament, the sergeant-at-arms, the directors or assistant directors of the main divisions of the parliamentary organization, and committee secretaries.

The self-assessment is based on a set of value judgments, with no right or wrong answers. The participants gave their ratings individually and anonymously, with discussion among themselves if they wished. A significant problem that surfaced was that different members of the group understood certain questions differently. The group thought that these questions were inherently ambiguous.

Several participants suggested questions that were not in the printed list and then proceeded to give ratings for those questions. These additions tended to be in areas of individual specialty, which may explain why, in most cases, no more than one person suggested a question in a particular area. Some broader areas of concern not covered by the printed questions were also raised. Results of the self-assessment are shown below:

Question	Rating
1. The representativeness of parliament	
1.1 How adequately does the composition of parliament represent the diversity of political opinion in the country (for example, as reflected in votes for the respective political parties)?	4
1.2 How representative of women is the composition of parliament?	2
1.3 How representative of marginalized groups and regions is the composition of parliament?	3
1.4 How easy is it for a person of average means to be elected to parliament?	1–2
1.5 How adequate are internal party arrangements for improving imbalances in parliamentary representation?	2
1.6 How adequate are arrangements for ensuring that opposition and minority parties or groups and their members can effectively contribute to the work of parliament?	4
1.7 How conducive is the infrastructure of parliament, and its unwritten mores, to the participation of women and men?	3
1.8 How secure is the right of all members to freely express their opinions, and how well are members protected from executive or legal interference?	4
1.9 How effective is parliament as a forum for debate on questions of public concern?	4–5
1.10 Additional questions:	
1.10.1 How effective in ensuring representativeness is the proportional representation system being used at present?	1
1.10.2 How adequately is the opposition resourced to carry out its functions?[1]	3

What has been the biggest recent improvement in the above?

- The Select Committee on Electoral Reform was appointed.
- Buddhist monks were elected to parliament.[2]

What is the most serious ongoing deficiency?

- Members are not responsible for a particular constituency.
- Weak opposition exists.

Benchmarking and Self-Assessment for Parliaments • http://dx.doi.org/10.1596/978-1-4648-0327-7

- Statesmanship is lacking.
- Only those rich enough can be elected.
- Nominations are given to kith and kin while educated people are reluctant to join the political process.
- Representation of women is low.

What measures would be needed to remedy this deficiency?

- Constitutional reform is necessary.
- Electoral reform is necessary.
- The political process must be cleaned up.
- Parties must encourage the greater participation of women.

Author's comments:

- Sri Lanka's present electoral system is based on proportional representation. The general feeling in the country appears to be that the first-past-the-post system produced more "representativeness," with a more meaningful relationship between parliamentarians and their constituents. The Select Committee on Electoral Reform, which has been sitting over a long period, spanning the life of two parliaments, was set up to address this situation.
- Women's representation in parliament is very low, the proportion being below the global and Commonwealth averages shown on the Inter-Parliamentary Union and Commonwealth Parliamentary Association websites. Interestingly, however, women parliamentarians in Sri Lanka have, over the life of many parliaments, ended up holding a large number of high posts, making the proportion of women in such positions higher than in most comparable countries. Most observers will agree that this is not tokenism—indeed that view would be difficult to maintain in a country where the posts of president and prime minister have been held by women, who are also well represented in professions such as the judiciary, university education, medicine, and engineering.

2. *Parliamentary oversight of the executive*

2.1 How rigorous and systematic are the procedures whereby members can question the executive and secure adequate information from it?   3

2.2 How effective are specialist committees in carrying out their oversight function?   3

2.3 How well is parliament able to influence and scrutinize the national budget through all its stages?   3–4

2.4 How effectively can parliament scrutinize appointments to executive posts and hold their occupants to account?   3

2.5	How far is parliament able to hold nonelected public bodies to account?	3
2.6	How far is parliament autonomous in practice from the executive (for example, through control over its own budget, agenda, timetable, personnel, and so forth)?	3
2.7	How adequate are the numbers and expertise of the professional staff to support members, individually and collectively, in the effective performance of their duties?	3–4
2.8	How adequate are the research, information, and other facilities available to all members and their groups?	3
2.9	Additional question:	
2.9.1	How far do the directives of oversight committees carry binding authority?	2

What has been the biggest recent improvement in the above?

- The Public Accounts Committee (PAC) and Committee on Public Enterprises (COPE) are now being assisted by experts.
- The 17th amendment to the constitution gives powers to a Constitutional Council; the budget is discussed in committees at length.
- The PAC and COPE Strengthening Project is being funded by World Bank.
- Some improvements in research work for oversight committees have occurred, but more strengthening is necessary.
- The Prebudget Select Committee gives more parliamentary control over finance.

What is the most serious ongoing deficiency?

- No follow-up on committee recommendations has occurred.
- The Constitutional Council is not functioning at the moment.
- The president holds many portfolios.
- A lack of interest among members of parliament sometimes leads to difficulty finding a quorum.
- Prebudget review is not sufficient.

What measures would be needed to remedy this deficiency?

- Standing orders need amendment.
- The opposition must stand against parliament losing control of funds—some individuals have even sought judicial intervention.
- Selection of members of oversight committees should be from those with necessary knowledge and interest.
- An overall change in approach and attitudes is needed.
- A committee of review (postbudget) should be set up.

3. Parliament's legislative capacity

3.1 How satisfactory are the procedures for subjecting draft legislation to full and open debate in parliament?	5
3.2 How effective are committee procedures for scrutinizing and amending draft legislation?	4
3.3 How systematic and transparent are the procedures for consultation with relevant groups and interests in the course of legislation?[3]	No consensus
3.4 How adequate are the opportunities for individual members to introduce draft legislation?	4
3.5 How effective is parliament in ensuring that legislation enacted is clear, concise, and intelligible?[4]	No consensus
3.6 How careful is parliament in ensuring that legislation enacted is consistent with the constitution and the human rights of the population?	4
3.7 How careful is parliament in ensuring a gender-equality perspective in its work?	3
3.8 Additional questions:	
3.8.1 How satisfactory are the safeguards with regard to the formulation of secondary or delegated legislation?	1
3.8.2 How far does the executive interfere with the work of the legislature?[5]	4

What has been the biggest recent improvement in the above?

- No responses were given.

What is the most serious ongoing deficiency?

- No detailed scrutiny of government bills is provided by committees.
- Debates in parliament do not focus on the core issues.
- Participation of members of parliaments in committees is poor: they do not read bills or materials supplied to them.
- The executive has taken over legislative activities.
- Time for discussion of urgent bills is lacking.
- Parliamentarians need to be educated about their duties and responsibilities.

What measures would be needed to remedy this deficiency?

- All bills should be referred to legislative committees.
- Review of subordinate legislation is necessary.
- The fundamentals of democracy need to be implemented.
- More time should be provided for public scrutiny of urgent bills.
- More training should be provided for both parliamentarians and their staff.

Author comment:

- The constitution provides for expeditious handling of a bill that the Cabinet has deemed "urgent." In effect this provision has meant that such a bill passes through its various stages very quickly and with hardly any discussion. As a regular practice, government bills are referred to a committee of the whole house, where passage is quicker but examination is less detailed.

4. The transparency and accessibility of parliament

4.1 How open and accessible to the media and the public are the proceedings of parliament and its committees?	3
4.2 How free from restrictions are journalists in reporting on parliament and the activities of its members?	3
4.3 How effective is parliament in informing the public about its work through a variety of channels?	2
4.4 How extensive and successful are attempts to interest young people in the work of parliament?	No consensus
4.5 How adequate are the opportunities for electors to express their views and concerns directly to their representatives, regardless of party affiliation?	No consensus
4.6 How user-friendly is the procedure for individuals and groups to make submissions to a parliamentary committee or commission of inquiry?	4
4.7 How much opportunity do citizens have for direct involvement in legislation (for example, through citizens' initiatives and referenda)?	2

What has been the biggest recent improvement in the above?

- Development of the parliamentary website has helped.
- Appointment of opposition members as chairs of PAC and COPE has helped (though not a regular occurrence).

What is the most serious ongoing deficiency?

- Committee work in camera is seriously lacking *(comment by several).*
- Accessibility is restricted at present by security considerations *(comment by several).*

What measures would be needed to remedy this deficiency?

- Telecast all proceedings.

Author comments:

- Schoolchildren are commonly seen going through the various open areas and in the gallery of parliament even with the very strict security of recent times. However, no planned program exists for young people to get involved in what parliament stands for and how it works. Attempts have been made to hold youth parliaments in live situations and on television, but these initiatives have never reached their expected fruition.

- Similarly, the scope for parliamentary education programs for the populace in general exists but remains unfulfilled.
- A parliamentary website has been set up and recently revamped in a more user-friendly format. This initiative will undoubtedly help take the parliament to the people.
- Although any person may be present in the gallery for plenary sessions, including meetings of committees of the whole house, meetings of committees are not open to the public.

5. *The accountability of parliament*

5.1 How systematic are arrangements for members to report to their constituents about their performance in office?	2
5.2 How effective is the electoral system in ensuring the accountability of parliament, individually and collectively, to the electorate?	2
5.3 How effective is the system for ensuring the observance of agreed codes of conduct by members?	2
5.4 How transparent and robust are the procedures for preventing conflicts of financial and other interest in the conduct of parliamentary business?	2
5.5 How adequate is the oversight of party and candidate funding to ensure that members preserve independence in the performance of their duties?	1
5.6 How publicly acceptable is the system whereby members' salaries are determined?	2
5.7 How systematic are the monitoring and review of levels of public confidence in parliament?	2

What has been the biggest recent improvement in the above?
- Discussions take place regarding the live telecast of proceedings in parliament.

What is the most serious ongoing deficiency?
- The present electoral system does not contribute to the accountability of members of parliament in the fullest sense.
- The implementation process (regarding telecasts and the like) is very slow.
- The salaries of members of parliament are too high (they are related to those of the judiciary).
- Some members of parliament do not make asset declarations.

What measures would be needed to remedy this deficiency?
- The present electoral system should be changed.

- Pressure from professional organizations and social groups is needed.
- Members' salaries should be on a par with the public service (which are in general lower than those of the judiciary).
- Asset declarations should be mandatory, and there should be a code of conduct.

Author comments:

- A uniformly low set of ratings for this area is indicative of a perceived lack of accountability and transparency with regard to members of parliament and their mores. Asset declarations are mandatory under law, but the practice has not been enforced. The number of members submitting declarations has improved considerably in the recent past.
- Members' salaries cannot be regarded as being high by the standards of most parliaments.

6. *Parliament's involvement in international policy*

6.1	How effectively can parliament scrutinize and contribute to the government's foreign policy?	3
6.2	How adequate and timely is the information available to parliament about the government's negotiating positions in regional and international bodies?	3
6.3	How much can parliament influence the binding legal or financial commitments made by the government in international forums, such as the United Nations?	3
6.4	How effective is parliament in ensuring that international commitments are implemented at the national level?	2–3
6.5	How effectively can parliament scrutinize and contribute to national reports to international monitoring mechanisms and ensure follow-up on their recommendations?	1–2
6.6	How effective is parliamentary monitoring of the government's development policy, whether as "donor" or "recipient" of international development aid?	3
6.7	How rigorous is parliamentary oversight of the deployment of the country's armed forces abroad?	No consensus
6.8	How active is parliament in fostering political dialogue for conflict resolution, both at home and abroad?	3
6.9	How effective is parliament in interparliamentary cooperation at the regional and global levels?	4
6.10	How much can parliament scrutinize the policies and performance of international organizations such as the United Nations, the World Bank, and the International Monetary Fund, to which its government contributes financial, human, and material resources?	3

What has been the biggest recent improvement in the above?
- Parliament introduced new legislation ensuring compliance with United Nations conventions and agreements.

What is the most serious ongoing deficiency?
- These are subjects handled mainly by the executive.

What measures would be needed to remedy this deficiency?
- Constitutional amendments are needed to give more powers to the legislature.
- All forums available in parliament should be used to discuss these issues (for example, consultative committees and parliamentary associations).

Although assessment efforts of this nature, even if numeral based, are meant to generate discussion rather than to be simple mathematical rating exercises, many lessons may be learned from a study of the ratings themselves. For instance, they may draw attention to the various ways in which different participants understand a question (and ambiguity may not be the sole cause of such different perceptions), and wide variations in the response of a group of participants could draw attention to situations of great significance to the study. Some amendments suggested to the IPU toolkit as a result of this exercise are as follows:

- The questionnaire does not cover adequately the situation of a presidential-parliamentary system such as that in Sri Lanka.
- Questions 2.1 and 3.1 are ambiguous or misstated—to the extent that procedures may exist—but what is important as a gauge of effectiveness is whether they are used in the intended manner.
- Regarding question 4.1, a very real problem in the wording was pointed out in that in the Sri Lankan parliament and many others—especially those deriving their existence from colonial legislatures—the public and the press may attend any plenary session but not those of committees.
- Another weakness that exists in many parliaments, including that of Sri Lanka, is their lack of control over delegated or subordinate legislation. This situation weakens the legislature and strengthens the executive. The questionnaire should perhaps address this point under section 3.

CPA Benchmarks and the Sri Lankan Parliament

The CPA benchmarks represent a list of 87 best practices and guidelines for self-assessment by democratic parliaments. Like the IPU toolkit, the CPA benchmarks are based on self-assessment. The benchmarks first gauge the general structure of the parliament, next observe the organization of the parliament, then look at its functionality, and finally delve into the underlying value system of parliament. See chapter 3 of this volume for further details.

The assessment, which follows, is based on interviews with a large number of participants ranging from ministers and members of parliament (from the government and opposition) to officials including the secretary general of parliament, his deputy, and the auditor general:

I. General
   1.1 Elections

Criteria:

   1.1.1 Members of the popularly elected or only house shall be elected by direct universal and equal suffrage in a free and secret ballot.
   1.1.2 Legislative elections shall meet international standards for genuine and transparent elections.
   1.1.3 Term lengths for members of the popular house shall reflect the need for accountability through regular and periodic legislative elections.

Assessment:

The main requirement for an elector is that he or she be a Sri Lankan citizen at least 18 years of age. Suffrage has been universal and equal in Sri Lanka since 1931, making its citizens among the world's earliest to enjoy that privilege. According to articles 88 and 89 of the constitution, the same eligibility criteria apply to those standing for election, with further provisions that they shall not be holders of public office. Voting is carried out under strict conditions of secrecy; attempts at multiple voting are not uncommon, but finger marking has proved a good deterrent.

Observer groups have described parliamentary elections as well administered and meeting the conditions for classification as "overall free and fair." However, several recommendations have been made on how to improve the process and remove advantages available to the government in power. A major disruption to the electoral process took place at the 2005 presidential election when the rebel LTTE prevented a large portion of the population from exercising its right to a free vote. Intimidation by armed groups at the provincial level has also been criticized.

The term of parliament is limited to six years. If parliament is not dissolved by presidential proclamation before that time, an automatic dissolution comes into force. Conversely, parliament may not be dissolved in the first year of its existence even if a government has been voted out by a no-confidence measure. However, under these circumstances, parliament may vote to dissolve itself.

   1.2 Candidate Eligibility

Criteria:

   1.2.1 Restrictions on candidate eligibility shall not be based on religion, gender, ethnicity, race, or disability.

1.2.2 Special measures to encourage the political participation of marginalized groups shall be narrowly drawn to accomplish precisely defined, and time-limited, objectives.

Assessment:

Sri Lanka's constitution and other legislation affecting elections do not contain any of the restrictions on candidate eligibility that are mentioned in the CPA benchmarks. Moreover, no affirmative action measure encourages the participation of minority and marginalized groups. Minority participation in Sri Lanka's political life is fairly standard, particularly through opportunities given by the major political parties (whose manifestos are not usually based on meeting communal or sectarian objectives).

Women's participation has always been a reality and is reflected in the political activity of the country. The extent of participation by women at parliamentary level remains numerically small, at about 8 percent of members, but this low participation appears to be a matter of choice. The figure is somewhat greater in local government. However, women who enter parliament usually obtain nomination to high posts, and the proportion of women members of parliament holding such posts is therefore higher than in most countries.

1.3 Incompatibility of Office

Criteria:

1.3.1 No elected member shall be required to take a religious oath against his or her conscience in order to take his or her seat in the legislature.

1.3.2 In a bicameral legislature, a legislator may not be a member of both houses.

1.3.3 A legislator may not simultaneously serve in the judicial branch or as a civil servant of the executive branch.

Assessment:

Members of Parliament (MPs) are given the option of taking an oath or making an affirmation according to their beliefs before they take their seat in the legislature.

Holders of public office, in any branch of the state, are not permitted to serve as MPs.

1.4 Immunity

Criteria:

1.4.1 Legislators shall have immunity for anything said in the course of the proceedings of legislature.

1.4.2 Parliamentary immunity shall not extend beyond the term of office, but a former legislator shall continue to enjoy protection for his or her term of office.

1.4.3 The executive branch shall have no right or power to lift the immunity of a legislator.

1.4.4 Legislators must be able to carry out their legislative and constitutional functions in accordance with the constitution, free from interference.

Assessment:

A legislator has immunity for anything said or done in the house during, but not beyond, his or her term of office. The speaker must act if a member abuses the privileges and immunities of the house. The courts have historically recognized the supremacy of parliament, and many judges have refused to intervene in cases brought before them over matters arising from speech, debate, or proceedings in the house on the basis that the courts had no jurisdiction in those circumstances.

A disturbance to this relationship arose in 2001, when the Supreme Court issued a stay order restraining the speaker from appointing a parliamentary select committee to inquire into the conduct of the chief justice following a motion of impeachment against him. In a well-researched and forthright defense of parliament's privileges and powers, Anura Bandaranaike, who was then the speaker, issued a landmark ruling that the courts had no jurisdiction to issue the stay orders on him and that he was therefore instructing the secretary general of parliament to place the motion on the order paper.[6] Very recently, a difference of opinion about the ability of the judiciary to rule on the operations of a parliamentary select committee arose, once again involving the impeachment of a chief justice, and once again parliament has asserted its independence.

1.5 Remuneration and Benefits

Criteria:

1.5.1 The legislature shall provide proper remuneration and reimbursement of parliamentary expenses to legislators for their service, and all forms of compensation shall be allocated on a nonpartisan basis.

Assessment:

All MPs of a particular rank or position are remunerated equally, irrespective of their party affiliation. Parliament no longer directly legislates on this matter for itself, because the levels of payment have recently been equated to those in the judiciary.

1.6 Resignation

Criteria:

1.6.1 Legislators shall have the right to resign their seats.

Assessment:

Article 66(b) of the constitution provides for resignation as one of the means by which a legislator may leave office.

1.7 Infrastructure

Criteria:

> 1.7.1 The legislature shall have adequate physical infrastructure to enable members and staff to fulfill their responsibilities.

Assessment:
Sri Lanka's parliament enjoys the use of a purpose-built complex outside the city of Colombo. Generous facilities are provided for members of parliament holding positions of authority in the house. Staff members are also adequately catered for. Members of parliament who live outside Colombo and the vicinity of parliament are provided with housing in official quarters close to parliament.

II. Organization of the Legislature
2. Procedure and Sessions
  2.1 Rules of Procedure

Criteria:

> 2.1.1 Only the legislature may adopt and amend its rules of procedure.

Assessment:
The constitution grants to parliament the authority to make its own rules of procedure subject only to such requirements as the election of a speaker. Also, according to the constitution, the Sri Lankan president has the authority to summon, prorogue, and dissolve parliament, but parliament decides on its own timetable for sittings and adjournments irrespective of their length (provided that a sitting is held at least once a year).

2.2 Presiding Officers

Criteria:

> 2.2.1 The legislature shall select or elect presiding officers pursuant to criteria and procedures clearly defined in the rules of procedure.

Assessment:
Although the requirement to elect presiding officers is specified in the constitution, standing orders 4 and 6 lay out the method of election. The previous speaker of parliament was elected in 2004 as a nominee of an opposition party. He was chosen in a contested election, but in other instances a member of the opposition has been elected unanimously to the post of speaker.

2.3 Convening Sessions

Criteria:

> 2.3.1 The legislature shall meet regularly, at intervals sufficient to fulfill its responsibilities.

2.3.2 The legislature shall have procedures for calling itself into regular session.

2.3.3 The legislature shall have procedures for calling itself into extraordinary or special session.

2.3.4 Provisions for the executive branch to convene a special session of the legislature shall be clearly specified.

Assessment:

The Sri Lankan parliament meets from Tuesday to Friday every other week during the year; the speaker, in conjunction with party leaders and whips, must determine any exceptions in advance. Apart from the day-to-day adjournment, no other type of vacation period is observed.

The days and hours of meetings are specified in the standing orders but may be varied with the permission of the house, which is usually achieved by prior agreement among the political parties (in effect, by leaders and whips). This arrangement results in a great deal of flexibility: the house may transact business well beyond the usual hours or meet on a "nonsitting" day.

Parliament thus has considerable freedom in determining meeting dates and times. The speaker may also call parliament in for an extraordinary session. An extraordinary session will be called only at a time of national emergency and must be initiated by a request of the prime minister (standing order 14). In addition, the president may call a special session of parliament during a period of prorogation or even recall, after dissolution but before the holding of a general election, the parliament that existed before the dissolution (Constitution of Sri Lanka, article 70).

2.4 Agenda

Criteria:

2.4.1 Legislators shall have the right to vote to amend the proposed agenda for debate.

2.4.2 Legislators in the lower or only house shall have the right to initiate legislation and to offer amendments to proposed legislation.

2.4.3 The legislature shall give legislators adequate advance notice of session meetings and the agenda for the meeting.

Assessment:

The Sri Lankan parliament has established a committee that coordinates parliamentary business and, in common with other committees, all parties are represented on that committee. The matters for discussion are then placed on the order paper, and the secretary general makes copies available to all members.

The standing orders require that a minister place certain bills before parliament. The most important of these bills are those dealing with finance and taxation. Any member may move an amendment or even present a bill, provided that it does not increase taxation or impose charges on the consolidated fund. The private member presenting a bill or amendment may seek the advice and assistance of the secretary general to ensure that the bill conforms to parliamentary standards.

2.5 Debate

Criteria:

2.5.1 The legislature shall establish and follow clear procedures for structuring debate and determining the order of precedence of motions tabled by members.

2.5.2 The legislature shall provide adequate opportunity for legislators to debate bills prior to a vote.

Assessment:

Standing orders 19 and 20 outline the usual order of business in the house, and standing orders 78 to 85 specify the rules of debate. The procedures to be observed by the presiding officer and members carrying out business are also discussed in detail.

Members may speak in any of the three languages recognized by parliament (Sinhala, Tamil, and English), and simultaneous interpretation into the other languages is provided.

2.6 Voting

Criteria:

2.6.1 Plenary votes in the legislature shall be public.

2.6.2 Members in a minority on a vote shall be able to demand a recorded vote.

2.6.3 Only legislators may vote on issues before the legislature.

Assessment:

Voting in parliament, whether it meets in plenary session or as a committee of the whole house, is public. Any member is free to demand a recorded vote, which is carried out by ringing the division bells and then taking a roll call. To vote, a member must be physically present in the chamber at his or her allotted seat.

2.7 Records

Criteria:

2.7.1 The legislature shall maintain and publish readily accessible records of its proceedings.

Assessment:

The secretary general is required to maintain the minutes of each day's proceedings and to prepare and print an official report (known, as in most Commonwealth countries, as *Hansard*) for the reference and use of members and the public (standing order 9). Parliament may make electronic or other records of proceedings, but these records have no official recognition or status.

3. Committees

3.1 Organization and 3.2 Powers

Criteria:

    3.1.1 The legislature shall have the right to form permanent and temporary committees.

    3.1.2 The legislature's assignment of committee members on each committee shall include both majority and minority party members and reflect the political composition of the legislature.

    3.1.3 The legislature shall establish and follow a transparent method for selecting or electing the chairs of committees.

    3.1.4 Committee hearings shall be in public. Any exceptions shall be clearly defined and provided for in the rules of procedure.

    3.1.5 Votes of committee shall be in public. Any exceptions shall be clearly defined and provided for in the rules of procedure.

    3.2.1 There shall be a presumption that the legislature will refer legislation to a committee, and any exceptions must be transparent, narrowly defined, and extraordinary in nature.

    3.2.2 Committees shall scrutinize legislation referred to them and have the power to recommend amendments or amend the legislation.

    3.2.3 Committees shall have the right to consult and/or employ experts.

    3.2.4 Committees shall have the power to summon persons, papers and records, and this power shall extend to witnesses and evidence from the executive branch, including officials.

    3.2.5 Only legislators appointed to the committee, or authorized substitutes, shall have the right to vote in committee.

    3.2.6 Legislation shall protect informants and witnesses presenting relevant information to commissions of inquiry about corruption or unlawful activity.

Assessment:

Sri Lanka's parliament relies on member committees to carry out a large part of its work. These committees are not referred to in the constitution, but they are set up as necessary under the standing orders or by resolution of parliament, and include a committee of the whole parliament, select committees, consultative committees, standing committees, and committees for special purposes.

A committee of the whole must be established to consider the clauses of an appropriation bill. A committee of the whole generally uses the same rules of procedure as parliament.

Select committees are ad hoc entities appointed to inquire into specific matters. The speaker appoints the chair and members of select committees, whereas in other committees the membership is determined by the Committee of Selection. Moreover, a select committee continues its work even if parliament is adjourned, it may resume with the same membership after a prorogation, and it is not dissolved until it reports back to parliament.

Consultative committees are set up to examine the work of each ministry allotted to a cabinet minister. The membership of each committee must reflect the composition of the house, with nominations being in the hands of the Committee of Selection and the chair as the respective cabinet minister.

Standing committees are responsible for examining bills referred to them by the house and for reporting back to the house. Membership is determined by the Committee of Selection, and the chair is elected by the members of the committee. Although a standing committee may continue its deliberations through an adjournment, a new committee is appointed after a prorogation, and the previous proceedings are referred to that committee.

The Committee of Selection is appointed at the beginning of every session. It consists of the speaker as chair and a number of members (including the leaders of all political parties in parliament or their nominees) specified from time to time by standing orders. This committee provides a consultation mechanism for determining the membership of all other committees. Once appointed, each member functions in a personal capacity, and neither the Committee of Selection nor party leaders may remove members.

Committees for special purposes (standing orders 121 to 126) include, among others, the Committee of Selection, the Committee on Parliamentary Business, PAC, COPE, the Committee on Privileges, and the Committee on High Posts. The chair of each special-purpose committee is elected by the membership.

Committees may resolve themselves into subcommittees in the interests of efficiency. The quorum is fixed at 4 for all committees (except for a committee of the whole, for which the quorum is 20, and for legislative standing committees, where the maximum quorum is 7).

Furthermore, except in the case of a committee of the whole house, committee deliberations are not held in public, though experts and other witnesses may be summoned, and demands for papers and records may be made. Such witnesses are provided the same immunity as members and officials of parliament with regard to their evidence; interference with them will constitute a breach of privilege of parliament. Of course, once a committee report is presented to parliament, it is also made available to the media and the public.

4. Political Parties, Party Groups, and Cross-Party Groups

4.1 Political Parties

Criteria:

>   4.1.1 The right of freedom of association shall exist for legislators, as for all people.

4.1.2 Any restrictions on the legality of political parties shall be narrowly drawn in law and shall be consistent with the International Covenant on Civil and Political Rights.

Assessment:

The constitution guarantees freedom of association to all to legislators, but it also specifies that no political party or other association shall aim to establish a separate state within the territory of Sri Lanka (article 157A).

Conditions for forming political parties are spelled out in the Parliamentary Elections Act 1981 and are fairly easy to satisfy. At parliamentary elections held in 2004, 52 parties participated and 14 obtained at least one seat. At the most recent elections, held in 2010, most parties formed themselves into alliances, and four of these comprising 18 parties between them are represented in parliament. Eight other parties and several independent candidates failed to obtain seats.

4.2 Party Groups

Criteria:

4.2.1 Criteria for the formation of parliamentary party groups, and their rights and responsibilities in the legislature, shall be clearly stated in the Rules.

4.2.2 The legislature shall provide adequate resources and facilities for party groups pursuant to a clear and transparent formula that does not unduly advantage the majority party.

Assessment:

All parties with members elected to parliament are provided with facilities in parliament as may be required for group, caucus, and similar meetings. Cross-party groups may be formed.

4.3 Cross-Party Groups

Criteria:

4.3.1 Legislators shall have the right to form interest caucuses around issues of common concern.

Assessment:

Currently, the only cross-party groups in existence are a parliamentary women's group and country-related friendship groups. Members take more interest in the Sri Lankan branches of international organizations such as the CPA, the IPU, and the South Asian Association for Regional Cooperation Parliamentary Group. Delegations sent to the meetings of these organizations are generally representative of the parties or alliances in parliament.

5. Parliamentary Staff

5.1 General, 5.2 Recruitment, 5.3 Promotion, and 5.4 Organization and Management

Criteria:

5.1.1 The legislature shall have an adequate nonpartisan professional staff to support its operations, including the operations of its committees.

5.1.2 The legislature, rather than the executive branch, shall control the parliamentary service and determine the terms of employment.

5.1.3 The legislature shall draw and maintain a clear distinction between partisan and nonpartisan staff.

5.1.4 Members and staff of the legislature shall have access to sufficient research, library, and information, communication, and technology facilities.

5.2.1 The legislature shall have adequate resources to recruit staff sufficient to fulfill its responsibilities. The rates of pay shall be broadly comparable to those in the public service.

5.2.2 The legislature shall not discriminate in its recruitment of staff on the basis of race, ethnicity, religion, gender, disability, or, in the case of nonpartisan staff, party affiliation.

5.3.1 Recruitment and promotion of nonpartisan staff shall be on the basis of merit and equal opportunity.

5.4.1 The head of the parliamentary service shall have a form of protected status to prevent undue political pressure.

5.4.2 Legislatures should, either by legislation or resolution, establish corporate bodies responsible for providing services and funding entitlements for parliamentary purposes and providing for governance of the parliamentary service.

5.4.3 All staff shall be subject to a code of conduct.

Assessment:

The constitution provides that the secretary general of parliament appoint all necessary staff members with the approval of the speaker (article 65). This arrangement requires cooperation across parliament and treasury.

The secretary general is appointed by the president and is protected by the constitution from unfair removal from office. Many members of parliament are seconded from the public service for varying periods of time; however, while serving in parliament, they are under the jurisdiction of the secretary general. The legislature thus controls the parliamentary service and determines the terms of employment, which are based on public sector practice but generally offer more favorable amenities.

The total number of parliamentary staff members is approximately 850. The most senior staff members are the secretary general, the deputy secretary general, and the assistant secretary general, who take responsibility for the work of the chamber. There are 9 table officers and 15 library staff members, of whom 5 are research officers.

Parliamentary staff members have no separate code of conduct, but they are subject to the Establishments Code, which applies to the public service (except in the area of disciplinary action, for which a separate set of rules exists).

Parliamentarians may also employ a certain number of staff members, whose allowances are paid by the Ministry of Parliamentary Affairs but who are not members of the public service. The member concerned is responsible for their discipline, while parliament provides the usual facilities for their work.

Members have access to a reasonable library and to steadily expanding information technology facilities. On appointment, each member is provided with information technology facilities for his or her personal office outside parliament. Members can request assistance to back up their research, but the extent of help available is still inadequate, which has implications for carrying out of such functions as parliamentary oversight.

Parliament may accept offers of technical and development assistance from foreign and international organizations (often through cooperative action with departments of the executive). An ongoing project for the modernization of parliament may help rectify some of the shortcomings noted.

III. Functions of the Legislature

6. Legislative Function

6.1 General

Criteria:

6.1.1 The approval of the legislature is required for the passage of all legislation, including budgets.

6.1.2 Only the legislature shall be empowered to determine and approve the budget of the legislature.

6.1.3 The legislature shall have the power to enact resolutions or other nonbinding expressions of its will.

6.1.4 In bicameral systems, only a popularly elected house shall have the power to bring down government.

6.1.5 A chamber where a majority of members are not directly or indirectly elected may not indefinitely deny or reject a money bill.

Assessment:
As reported elsewhere in this chapter, the Sri Lankan parliament acts in accord with all of these benchmarks (excluding 6.1.4, which is not relevant in Sri Lanka's case).

6.2 Legislative Procedure

Criteria:

6.2.1 In a bicameral legislature there shall be clearly defined roles for each chamber in the passage of legislation.

6.2.2 The legislature shall have the right to override an executive veto.

Assessment:
Benchmark 6.2.1 is not relevant to the Sri Lankan parliament. With respect to benchmark 6.2.2, the constitution does not allow for an executive veto

on legislation. When a bill is passed by parliament, the speaker issues a certificate to that effect and the bill then becomes an act of parliament. In certain cases, such as a constitutional amendment, the certificate must be signed by the president, but there is no provision that allows either the president or the speaker to refuse to sign. Moreover, according to the constitution, the validity of the new act can no longer be questioned in a court of law (article 124).

6.3 The Public and Legislation

Criteria:

6.3.1 Opportunities shall be given for public input into the legislative process.

6.3.2 Information shall be provided to the public in a timely manner regarding matters under consideration by the legislature.

Assessment:

As reported elsewhere in this chapter, the Sri Lankan parliament acts in accord with both of these benchmarks.

7. Oversight Function

7.1 General

Criteria:

7.1.1 The legislature shall have mechanisms to obtain information from the executive branch sufficient to exercise its oversight function in a meaningful way.

7.1.2 The oversight authority of the legislature shall include meaningful oversight of the military security and intelligence services.

7.1.3 The oversight authority of the legislature shall include meaningful oversight of state-owned enterprises.

Assessment:

As mentioned earlier, the parliament of Sri Lanka has a large network of committees, and several of these were created to provide oversight in a meaningful way. Military and security matters, for instance, are examined by the consultative committee dealing with those functions.

7.2 Financial and Budget Oversight

Criteria:

7.2.1 The legislature shall have a reasonable period of time in which to review the proposed national budget.

7.2.2 Oversight committees shall provide meaningful opportunities for minority or opposition parties to engage in effective oversight of government expenditures. Typically, the public accounts committee will be chaired by a member of the opposition party.

7.2.3 Oversight committees shall have access to records of executive branch accounts and related documentation sufficient to be able to meaningfully review the accuracy of executive branch reporting on its revenues and expenditures.

7.2.4 There shall be an independent, nonpartisan supreme or national audit office whose reports are tabled in the legislature in a timely manner.

7.2.5 The supreme or national audit office shall be provided with adequate resources and legal authority to conduct audits in a timely manner.

Assessment:

As mentioned earlier, the parliament of Sri Lanka has a large network of committees and several of these are created with the intention of exercising its oversight function in a meaningful way. Military and security matters could, for instance, be examined by the consultative committee dealing with those functions.

There is no finance committee in the sense in which that term is generally used. The expectation would be that budgetary policy and other financial issues would be debated in the Finance Consultative Committee. Such debate has not happened in practice. The basic reason for this is not difficult to gauge. With 50-plus consultative committees and the more usual number of select and standing committees needed by any parliament to carry on its work efficiently, and with each committee consisting of 31 members, the energies of the 225 members of parliament are quickly absorbed. Many committees cannot carry out their functions for want of a quorum at meetings. The parliamentary staff members attached to committees are similarly stretched, though officials of the relevant ministry offer some assistance. Given that the minister is the chair, it is a moot point whether the independence of parliament could not in time become the loser in the whole process.

The budget itself is presented, as is common practice, at a plenary session of parliament. Until recently, Sri Lanka had one of the longer periods of time allocated by any parliament for debate on the budget, but this period has been cut by mutual consent to two weeks. Details are examined by a committee of the whole. Plenty of opportunity exists for serious and substantive debate, but in recent times, perhaps over the past 15 to 20 years, the quality of debate has deteriorated—a conclusion that is not generally contested. Some analysts attribute this decline to the current system of proportional representation obtaining in the country, which, they claim, has resulted in a member of parliament no longer being personally responsible to his or her electorate or district. Others believe the proportional representation system is not being used in the way it should be and that the remedy lies in better selection of candidates by political parties.

The constitution provides for the appointment of an auditor general by the president on the recommendation of the Parliamentary Council, which comprises the speaker, the prime minister, and the leader of the opposition, among others. The holder of the post is protected by the constitution (article 153) and is described as an officer of parliament—that is, he or she is responsible to parliament though not appointed by the speaker or secretary general of parliament and is afforded the same immunities as an officer of parliament in carrying out his or her duties. These immunities extend to officials assisting the auditor general.

The department of the auditor general is the supreme audit agency for the country. To ensure its independence, it does not fall under a minister but is otherwise operated on the same lines as a department of government. The auditor general is empowered to call on other recognized audit personnel outside the department when such services are required.

Reports from the auditor general to parliament are based on the annual accounts prepared by each department or public enterprise and, once tabled, are open to the media and the public. The department does not have the staff or other resources required to audit every organization in depth, so the attention of parliament is drawn to the cases that need to be highlighted. Such judgment calls are not always easy to make. Although the holders of the post have been generally recognized for their independence, their reports have often been delayed because the bodies being audited have been slow in coming up with supplementary information required.

Under the arrangements for committees for special purposes prescribed in standing orders, parliament has established two committees for the examination of past expenditures. One is PAC, which deals with government departments, and the other is COPE. The duty of these committees, as defined in the standing orders of parliament, is to examine the accounts of each government department or public corporation in which the government has a controlling share, alongside the relevant report of the auditor general. Each committee is required to report to parliament from time to time on the accounts and finances, financial procedures, performance, and management of the institution. In practice, the committees are unable to examine every set of accounts and tend to concentrate on specific matters that are drawn to their attention by the auditor general.

Unfortunately, PAC and COPE reports often suffer from not being sent back expeditiously to parliament or not being debated thereafter in parliament with the enthusiasm they may deserve. Nevertheless, in some important cases, such a report has resulted in vigorous action against the person or institution concerned. At present, some highly critical reports have been presented to parliament. The appointment of chairs of committees follows regulations specified in the standing orders. Because PAC and COPE are special-purpose committees, the selection of the chairs is left to the membership of each committee. Given that committee membership reflects the strength of parties in parliament, the chair could be selected from among government members if committee membership so wishes.

At independence and in the years following, it had been the custom, in line with the traditions of Westminster, for PAC to be chaired by an opposition member. This practice changed in the 1960s, when certain opposition parties joined the government in a coalition. The member chairing PAC, who now found himself in government but was recognized by all sides as a very effective chair, was by general agreement permitted to continue in that office. Since then, the tradition that an opposition member should chair PAC has not always been observed, but several chairs from the government benches have shown themselves capable of preserving the independence of the post.

Opposition members who are dissatisfied with the handling of some part of PAC's (or COPE's) scrutiny function can always take up matters by submitting a dissenting report (a path that politicians tend to avoid), by questioning the minister concerned in a plenary session of parliament, or by having the item raised in the consultative committee (the latter would need either the agreement of the chair, who it must be noted is the minister, or a reference to the committee by parliament).

PAC and COPE have not so far worked with independent audit authorities but have relied on the auditor general and his or her staff. As mentioned earlier, the auditor general may obtain services from outside his or her department, but this step has been occasioned more by shortage of resources than by the need to obtain access to specialist skills.

Each minister is present when his or her ministry's estimates of expenditure are taken up for discussion, and the minister is therefore available to question about the operations of the ministry. It is a matter of some concern that questions raised tend to be of a simplistic and parochial nature.

7.3 No Confidence and Impeachment

Criteria:

    7.3.1 The legislature shall have mechanisms to impeach or censure officials of the executive branch or express no confidence in the government.

    7.3.2 If the legislature expresses no confidence in the government, the government is obliged to offer its resignation. If the head of state agrees that no other alternative government can be formed, a general election should be held.

Assessment:

The constitution provides parliament with specific powers to impeach the president. The Sri Lankan parliament has a specific set of mechanisms in place to impeach or censure officials of the executive branch or to express no confidence in the government. Motions of no confidence may be moved against an individual or against the cabinet as a whole. The censure of an individual is resolved through resignation of the person concerned and has possible implications for the government as a whole. A no-confidence motion in the cabinet always results in the government's resignation, and either the parliament must be dissolved by the president or the leader of another political party must be called on to form a new government.

Separate provisions apply that permit action to be initiated in parliament against judges of the Supreme Court and the appeal courts. Parliament may also resolve that officials whose positions are protected by the constitution (for instance, the secretary general of parliament or the auditor general) be removed from office by the president.

The constitution restricts the president's right to dissolve parliament if no confidence is expressed by the rejection of the statement of government policy at the first session following a general election. Parliament is guaranteed a minimum life of one year unless it resolves to ask the president to dissolve it.

8. Representational Function

8.1 Constituent Relations and 8.2 Parliamentary Networking and Diplomacy

Criteria:

8.1.1 The legislature shall provide all legislators with adequate and appropriate resources to enable the legislators to fulfill their constituency responsibilities.

8.2.1 The legislature shall have the right to receive development assistance to strengthen the institution of parliament.

8.2.2 Members and staff of parliament shall have the right to receive technical and advisory assistance, as well as to network and exchange experience with individuals from other legislatures.

Assessment:
These functions and responsibilities (benchmarks 8.1 and 8.2) are more fully discussed elsewhere (see, for instance, benchmarks 5.1 and 5.4).

IV. Values of the Legislature

9. Accessibility

9.1 Citizens and the Press

Criteria:

9.1.1 The legislature shall be accessible and open to citizens and the media, subject only to demonstrable public safety and work requirements.

9.1.2 The legislature should ensure that the media are given appropriate access to the proceedings of the legislature without compromising the proper functioning of the legislature and its rules of procedure.

9.1.3 The legislature shall have a nonpartisan media relations facility.

9.1.4 The legislature shall promote the public's understanding of the work of the legislature.

Assessment:
As reported elsewhere in this chapter, the Sri Lankan parliament acts in accord with all of these benchmarks.

9.2 Languages

Criteria:

> 9.2.1 Where the constitution or parliamentary rules provide for the use of multiple working languages, the legislature shall make every reasonable effort to provide for simultaneous interpretation of debates and translation of records.

Assessment:

As reported elsewhere in this chapter, the Sri Lankan parliament acts in accord with all of these benchmarks.

10. Ethical Governance
10.1 Transparency and Integrity

Criteria:

> 10.1.1 Legislators should maintain high standards of accountability, transparency, and responsibility in the conduct of all public and parliamentary matters.
>
> 10.1.2 The legislature shall approve and enforce a code of conduct, including rules on conflicts of interest and the acceptance of gifts.
>
> 10.1.3 Legislatures shall require legislators to fully and publicly disclose their financial assets and business interests.
>
> 10.1.4 There shall be mechanisms to prevent, detect, and bring to justice legislators and staff engaged in corrupt practices.

Assessment:

Members of parliament are not protected by the constitution, other laws, or the rules of parliamentary immunity from action being taken against them under the laws of the country. The speaker would always be informed of such action, and precautions have to be taken to ensure that the member of parliament is not prevented from functioning in his or her legislative capacity without grave cause.

Parliament does not impose a code of conduct other than in the specific situations relating to parliament, which are covered by the standing orders. Members of parliament are subject to the Declaration of Assets and Liabilities Act 1981, which requires a declaration to be made to the speaker on election. However, these declarations are not available to the public. Similarly, legislators and staff members are subject to anticorruption legislation, including the Bribery Act 1973 and the Offences against Public Property Act 1982.

Conclusion

The aim of this chapter was to examine Sri Lanka's parliament through the lens of both the IPU parliamentary indicators and the CPA benchmarks, with updated information for both evaluations in light of recent events in the country, most notably the end of civil war in 2009.

Benchmarking and Self-Assessment for Parliaments • http://dx.doi.org/10.1596/978-1-4648-0327-7

One of the overarching themes that is revealed from studying both assessment frameworks is that although Sri Lanka is still a developing country, its parliament is a mature institution. Therefore, many of the outstanding problems do not stem from a lack of experience in legislative procedures and practice but rather from a resource shortfall. Several areas of the infrastructure need strengthening, and these areas appear to be receiving attention (particularly since the end of the war with the LTTE). However, the feeling remains that the collective political will is weak in confronting some fundamental situations and will probably remain so until the effects of the war—direct and indirect—cease to be pervasive.

The impression one receives from this work with the parliamentary staff is that the Sri Lankan legislature is reasonably well placed to address and service the demands of a democracy. However, the parliament does not use fully the powers and the procedures at its disposal, perhaps more because of a lack of will than of a lack of understanding.

Participants faced several challenges in applying the benchmarks and indicators to the Sri Lankan case, and attention was drawn to the lack of benchmarks in certain important areas. Despite these difficulties, the general view of participants in the parliamentary self-assessment is that the benchmarks and indicators constitute an important advancement in helping legislatures conduct self-assessments. Some participants added that more widespread use of these and other assessment schemes will require considerable effort by organizations such as the CPA, the IPU, and the World Bank. A concerted and coordinated push is needed to entrench the kind of thinking required by self-assessments and to establish this exercise as a regular routine of parliamentary administration.

Viewing the possible success of the exercise as a technique for identifying priorities and means for strengthening parliament, as mentioned in the introductory material, it is significant that participants observed a need for constitutional and electoral reform. The executive is considered to be too strong and to encroach on the powers of parliament. The present system of proportional representation is not seen as working well. The possible need for amendment of standing orders has been raised.

It now becomes important to use these self-assessment techniques to gauge the views of Sri Lankan parliamentarians on these points. The early years of this century have not been an opportune time to carry out such exercises in Sri Lanka because of the country's preoccupation with the war with the LTTE, but one hopes that parliamentarians will now resolve to turn their attention to these matters. The two self-assessments could be a way of encouraging that approach.

Notes

1. Suggested as an amendment to 1.6.
2. Traditionally Buddhist monks, like most clergy, did not enter into active politics, but they have great influence because of their association with the majority religion and their role as guardians of the nation's culture. A group of monks have banded into a political party, and varying views have been expressed on their entry into parliament.

3. The lack of consensus reflects ad hoc approaches to consultation.

4. See the statements regarding insufficient review by legislative committees.

5. This rating reflects high interference.

6. Ruling on Supreme Court Stay Order seeking to restrain the Speaker from appointing a Select Committee, June 20, 2001, Hansard of the Parliament of Sri Lanka.

References

CPA (Commonwealth Parliamentary Association). 2006. "Recommended Benchmarks for Democratic Legislatures." CPA, London. http://wbi.worldbank.org/wbi/Data/wbi/wbicms/files/drupal-acquia/wbi/Recommended%20Benchmarks%20for%20Democratic%20Legislatures.pdf.

Gomez, Raja. 2008a. "Benchmarks for Democratic Legislatures: Sri Lanka—A Study of the Application of the CPA Benchmarks to the Sri Lankan Parliament." World Bank, Washington, DC.

———. 2008b. "Parliamentary Indicators (Financial and Budgetary): Sri Lanka." World Bank, Washington, DC.

———. 2009. "Evaluating Parliament: Application of the IPU Self-Assessment Toolkit to the Sri Lanka Parliament." World Bank, Washington, DC.

IPU (Inter-Parliamentary Union). 2008. "Evaluating Parliament: A Self-Assessment Toolkit for Parliaments." IPU, Geneva. http://www.ipu.org/pdf/publications/self-e.pdf.

CHAPTER 11

Building on the CPA Benchmarks to Establish a Parliamentary Accountability and Management Framework: The Case of Canada

Jill Anne Joseph

Introduction

Parliamentary benchmarks have become popular among many international parliamentary organizations in the past few years. Benchmarking is a process used to set goals for improvement by assessing how institutions compare with good practices. A variety of internationally recognized standards and control frameworks are commonly used to assess organizational performance or compliance. In the parliamentary context, interparliamentary assemblies and other international institutions have developed a wide variety of benchmarks in recent years to help nations pursue continued democratic improvements. Many of these benchmarks also touch on management and accountability issues, but not in the depth needed to ensure adequate performance in these areas.

How can benchmarking be used to renew and reinforce parliaments' accountability and validity? To remain relevant and effective in the 21st century, legislatures must not only maintain modern management practices and systems, but also be seen as doing so. Benchmarks can create a model for the accountability of parliaments to their citizens. Such a model can ensure that parliaments' basic roles and responsibilities to the people are met and can help build or rebuild the trust lacking in legislatures worldwide.

At the administrative level, a framework of benchmarks should be established, shared, and integrated into parliamentary management practices. Parliamentarians and their supporting administrations must relinquish the notion that their institutions are too independent or unique to have to meet the basic standards of management and accountability required of public office holders and institutions. Parliaments need a management and accountability framework—a set of benchmarks that establishes the services that parliaments require of their administrations to function well and, in turn, to provide information and service to the voting public.[1]

This chapter outlines the development of a parliamentary accountability and management framework, drawing primarily from a Canadian perspective and Canadian examples. The Parliament of Canada completed a self-assessment against the Commonwealth Parliamentary Association (CPA) benchmarks in 2009 and found revealing results in terms of identifying new goals for its democratic reform (CPA 2010).

The rest of this chapter is organized as follows: The first section discusses guidelines for benchmarking frameworks. The second section proposes benchmarks that can be used to assess core processes, products, and services of parliamentary activities in an accountability and management framework. The final section concludes.

Guidelines for Benchmarking Frameworks

For the past six years, Canada's Senate Administration has been using a set of benchmarks called the Management Accountability Framework (MAF) to assess its management processes. MAF was developed by Canada's Treasury Board Secretariat, the administrative branch that advises and supports the Treasury Board, which is a committee of cabinet ministers, in its role to ensure value for money and oversight of the financial management functions in government departments and agencies. All federal government departments and agencies in Canada are subject to MAF assessments yearly or, in the case of small institutions, triennially.[2] Although houses of parliament are not subject to MAF reviews, the Senate Administration uses MAF for self-assessment purposes.

Initially, Senate senior management did not wholeheartedly embrace use of MAF, a tool developed for and by the executive branch to improve parliament's management processes. However, the Senate's first attempt of the MAF exercise was deemed valuable in some respects and has led to its annual recurrence. To further increase the usefulness of the MAF exercise, senior Senate officials have suggested that it be adapted and expanded to make it more suitable to the parliamentary context.

Although there is room for further development, MAF offers a useful starting point for developing a new framework because it covers a range of management functions that assess the Senate as a public institution. In particular, MAF sets out 10 basic management functions that should be present in any modern institution, with weighted criteria to rate institutional maturity in each area.[3]

The Senate is not alone in seeking a framework of benchmarks suited to legislatures. Vivek K. Agnihotri, secretary general of the Council of States of India, contributed a forward-looking article to *The Parliamentarian* in which he discusses various total quality management models, including the European Foundation for Quality Management Excellence Model and the Malcolm Baldrige National Quality Award (Agnihotri 2010). What needs to be assessed in all of these models, including Canada's MAF, is similar and can be broadly summarized as follows:

- Leadership (governance, tone at the top, values, and ethics)
- Strategy and planning (including multiyear investments and project management)
- Risk management (including business continuity planning)
- Health of the workplace and workforce
- Partnerships
- Client-focused service (including processes, products, and services)
- Performance measurement and reporting

By adapting these broad categories to individual needs, each parliament can ensure the presence and assess the quality and completeness of a core set of products, processes, and services. Such a framework serves in several ways:

- Identifying shortcomings and setting a strategic course to improve services to parliamentarians and citizens
- Developing strategic partnerships domestically or abroad, either to achieve economies of scale in the provision of services or to facilitate capacity building
- Determining the priority needs of developing or underresourced parliaments so that donor funding can be directed more effectively

A benchmark framework may also include an objective rating system of graduated criteria that reflect levels and quality of services, products, and processes. Such a rating scale should not consist of value judgments, but rather descriptive assessment criteria against which a legislature can self-assess and rank itself. The evidence of a legislature's level of compliance with the criteria within a benchmark can be easily validated, making self-evaluation a much more objective exercise. Table 11.1 shows an extract from MAF that provides an example of such graduated criteria.

A benchmark framework could also include criteria assessing essential services to parliamentarians and the public, as well as parliamentarians' and management's accountability.

Table 11.1 Effectiveness of Corporate Risk Management

Requires immediate attention	Needs improvement	Acceptable	Strong
Accountability for managing key risks does not appear to be assigned to senior management.	Accountability for managing key risks appears to be inconsistently assigned to senior management.	Accountability for managing key risks has been assigned to senior management.	Accountability for managing key risks has been clearly assigned to senior management, and performance is assessed.

Source: Extracted from Management Accountability Framework Rounds VI–VII.
Note: The related rounds of MAF contained over three-dozen criteria with varying levels of maturity relating to risk management.

Identifying Core Products, Processes, and Services

The core processes, products, and services of parliamentary activities can be broadly grouped into three overarching areas: governance and management; parliamentary information and public outreach; and legislative, oversight, and procedural functions. All three areas are useful in developing benchmark categories for essential activities and services offered to stakeholders such as parliamentarians, the public, and the media. This section first proposes benchmarks for each of the three areas in a parliamentary accountability and management framework and then discusses how to devise an assessment scale.

Governance and Management

Governance and management is the first essential area of benchmarks for inclusion in a parliamentary accountability and management framework. The CPA has done some excellent work in this area that provides inspiration for a number of benchmarks. For instance, the CPA's (2006) study group report "Recommended Benchmarks for Democratic Legislatures" has benchmarks covering internal matters (such as the need for nonpartisan parliamentary staff and ethical governance).[4] The CPA also produced a study group report titled "The Administration and Financing of Parliament," which promotes the need for administrative independence. Administrative independence is "best achieved through the establishment of parliamentary corporate bodies [to provide] a responsive and accountable parliamentary service ... A parliamentary corporate body is responsible for determining the range and standards of service to be provided, securing a parliamentary budget, providing leadership and direction to the parliamentary service, and reporting to Parliament and the public on performance and stewardship" (CPA 2005, 3, 9).

Governance benchmarks should also ensure the existence of an independent, all-party board or committee of parliamentarians that is empowered to provide management oversight on behalf of all parliamentarians.[5] Such a committee or board requires a relatively high degree of competency and continuity. This board may be chaired by the speaker and should fulfill the following roles:

- Working with the clerk and senior management to set strategic direction and planning
- Reviewing the risk environment, including business continuity planning
- Setting the tone at the top (values and ethics)
- Ensuring adequate human and financial resources and their appropriate allocations
- Defining required services and service standards
- Measuring, monitoring, and publicly reporting performance
- Selecting the clerk on the basis of merit
- Maintaining involvement (consultations) in management evaluations, compensation, and succession planning
- Establishing written rules or regulations to support their roles and the application of the statutory framework

Benchmarks should also assess the adequacy or frequency of meetings, member continuity, member orientation and training, methods of selecting members, and adequate provision of management reports and information.

Complementing and overlapping on this list, the CPA study group concluded that the following conditions and relationships would ensure effective delegation for governance:

- A corporate body should have a clear understanding of its role in setting strategic priorities and monitoring of progress;
- Given that members of corporate bodies have other political and parliamentary commitments there needs to be a dedicated secretariat to support the corporate body;
- Corporate body meetings and decision making need to be informed by the right agenda and appropriate management information;
- An unambiguous and positive relationship between the Speaker, corporate body, and the Clerk built on the principles of openness, integrity, and accountability;
- As Accounting Officer the Clerk should have ultimate financial responsibility for the Legislature;
- Development of a competent parliamentary service that provides assurance to the corporate body that its decisions are fully implemented; and
- Establishment of relevant House committees in special subject areas, e.g. finance, catering, environmental issues. (CPA 2005, 9–10)

The principles laid out by the CPA study group underscore that detailed benchmarks are key to the management efforts of the parliamentary service and the secretariat under the direction of the legislature's clerk. To that effect, well-established management criteria set out in management frameworks such as MAF could be adapted to the legislative environment. The study group's report makes a number of recommendations on governance, financial independence, parliamentary service, and public accountability that would in themselves make excellent benchmarks for democratic legislatures, but they may have been viewed as beyond the scope of the benchmarking exercise undertaken in 2006. However, the study group went so far as to discuss the development of an accountability framework, in which it included the following mechanisms:

Internal

- Estimates/corporate plans/financial plans
- Compliance with best practice accounting standards
- Internal audit reports
- Corporate audit committee
- Compliance audits against general legal requirements
- Customer surveys
- Equal opportunities policies

External

- Annual reports
- Audited accounts
- External audit reports
- Information strategy
- Education program[s]
- Response to oral and written questions
- Attendance at Public Accounts Committee (CPA 2005, 12)

Although these accountability mechanisms are discussed in detail in the next subsection about parliamentary information and public outreach, they are also relevant to benchmarks addressing governance and management. For instance, regarding the existence of a corporate audit committee, benchmarks should be set to ensure that such committees are working toward full compliance with international and domestic financial and internal audit standards. As a secretariat or parliamentary service's primary "customers," parliamentarians should be engaged in periodic consultations regarding their satisfaction with the nature and standards of service they receive. Additional input can also be gained from consulting "customers" of the legislature, such as committee witnesses, the media, the general public, and secretariat employees. Part of the challenge in establishing these consultation practices, however, is that secretariats have a monopoly as service providers.

Parliamentary Information and Public Outreach

The second essential area of benchmarks for inclusion in a parliamentary accountability and management framework is parliamentary information and public outreach. Worldwide scandals such as fraudulent activities in public corporations, the collapse of financial systems, and negative audit findings—from which parliaments have not been exempt—all contribute to a lack of public trust. Rebuilding that trust has been aided in part by strengthened regulatory controls and, in the case of some parliaments, the advent of increased public accountability and regular, independent audits.

Canada's electoral context helps illustrate the problem. A 2011 article in the *Toronto Star* observes a worldwide trend of a low public opinion of politicians and suggests that, at least in Canada, a sense of resignation or complacency among citizens has led to the notion that a single vote does little to affect parliamentarians' decision making once in office (Brennan 2011). Voter turnout in Canada averaged around 75 percent during the latter half of the 20th century, but fell to 58.8 percent in 2006 and increased only slightly to 61.4 percent in 2011. The Manning Centre (2011, 4) found that three-quarters of Canadians do not feel that their interests are represented in Ottawa, and 35 percent feel that members of parliament mainly focus on furthering their personal and career interests. Improved information sharing may be the single most effective way to increase transparency and accountability and to dispel the public doubts and accusations that legislatures endorse cultures of secrecy and entitlement.

A closer look at the Canadian case suggests that the bottleneck in information sharing is not availability of the information, but rather a lack of awareness or ease of access. According to Philip Laundy, a former clerk assistant of Canada's House of Commons, a wealth of parliamentary documents and information is available to the public, but readership of these official publications is low (Laundy 1989, 129–35).[6] The same holds true for broadcasting: a 2006 viewership study of Canada's Cable Public Affairs Channel indicated that the average audience of Senate hearings, for example, ranges from only 300 to 2,000 viewers at any given moment (Canadian Media Research 2006). The advent of webcasting of hearings would not have assisted this statistic.

Greater efforts are thus needed to provide parliamentary information in as many formats as possible in a way that maximizes media and public engagement while ensuring cost-efficiency and cost-effectiveness. In particular, public outreach could be conducted through parliamentary newsletters, broadcasts, youth visits, public town hall meetings, and blogs. However, part of the challenge is that so many actors are involved in public outreach—namely institutionwide efforts to educate or inform the population, committee or issue-based outreach by groups of legislators, communication efforts by political party caucuses within parliament, and individual member efforts (NDI and UNDP 2004, 3).

Despite this complexity, there is a wide consensus that outreach activities have many positive benefits. According to the National Democratic Institute for International Affairs and the United Nations Development Programme (UNDP), such activities

- Strengthen image of parliament as open and democratic: inform citizens of accomplishments and goals;
- Demonstrate that the group is working to advance citizen interests: articulate policy stances;
- Establish relationships with media: inform constituents of goals and minimize severe reactions to policy changes;
- Promote informed policymaking: ensure that parliament is truly representative;
- Group can better decide what policies are priorities: group will be better informed on substance of policy;
- Identify what issues matter most to constituents/district: improve public image;
- Demonstrate the effectiveness of parliament and its important role in solving the country's problems;
- Enhance reputation of group: can be seen as responsive to public needs;
- Improve ability to identify trends and recurrent problems: build loyalty among constituents. (NDI and UNDP 2004, 5)

Public recognition of these benefits is growing, as is the public's expectation that parliaments and parliamentarians provide a further level of transparency regarding their work, the funds available to them, and how they spend those funds. In response to this demand, most parliaments now produce traditional parliamentary records, annual reports,[7] education aids for teachers, statistical reports, and member biographies. Parliamentary subscribers have also submitted information for the production of a series of Global Parliamentary Reports published by the UNDP and the Inter-Parliamentary Union in 2011 on the relationship between parliamentarians and citizens, including the ways in which parliamentarians are communicating with voters. The invitation to contribute to this analysis requested descriptions of "the main challenges that parliaments and politicians are facing now—in terms of public opinion, trust, and expectations—and how they are responding."[8]

Benchmarks can play an important role in promoting effective public outreach by rating the completeness and maturity of outreach activities. Benchmarks that assess parliamentary information sharing should evaluate not only the records themselves, but also the variety of their formats, their timeliness and quality, their searchability, and the legal environment within which the media must operate (Mendel 2005), all with an awareness about cost-effectiveness.

A useful starting point for assessing public outreach is the CPA benchmarks, which advocate, under benchmark 6.3, on the public and legislation, that "[o]pportunities shall be given for public input into the legislative process" and that "[i]nformation shall be provided to the public in a timely manner regarding matters under consideration by the legislature." Under benchmark 9.1, on citizens and the press, the benchmarks also require that the legislature "be accessible and open to citizens and the media," that the media be "given appropriate access to the proceedings," that the legislature "have a nonpartisan media relations facility," and that the legislature "shall promote the public's understanding of the work of the legislature." These benchmarks could be further developed with rating criteria based on various outreach activities. Moreover, under section 10, which addresses ethical governance, the benchmarks require a code of conduct with rules on conflicts of interest, acceptance of gifts, and public disclosure of financial assets and business interests. Still other accountability benchmarks are detailed in the accountability framework mechanisms prescribed in the CPA study group report and could include criteria to rate levels of maturity so that legislators may identify areas for continuous improvement.

Legislative, Oversight, and Procedural Functions

The third essential area of benchmarks for inclusion in a parliamentary accountability and management framework is legislative, oversight, and procedural functions. As parliamentarians come and go over the years, the secretariat or parliamentary service must provide continuity and corporate memory to successfully support legislators in fulfilling their legislative, executive oversight, representative, and corporate roles and responsibilities. Facilities such as adequately stocked libraries, office and electronic equipment, and a professional research and technical staff with the education, knowledge, information, and independence to provide legislative and policy advice are basic requirements for parliamentarians.

Benchmarks can play an important role in assessing these functions. One source for guidelines on such benchmarks is the CPA's (2006) "Recommended Benchmarks for Democratic Legislatures,"[9] which focuses on ensuring that certain basic parliamentary legislative and oversight functions are in place.

However, these benchmarks may be too basic to evaluate whether a legislature is adequately supported to perform its roles well. Other useful guidance for benchmarks was developed by the CPA and the World Bank in a study group hosted by the Ontario legislature (McGee 2002):[10]

- Public accounts committees (PACs) should have frameworks of powers and practices:
  - The power to require the government to respond (this should be a power for all committees)
  - The power to sit outside session whenever necessary
  - A permanent order of reference to examine public accounts and legislative auditor's reports
- PACs should hold their meetings in public with full verbatim transcripts.
- PACs should hold deliberative meetings on their reports in camera.
- PACs should be chaired by an opposition member.
- PACs should avoid partisanship and seek consensus.

Several other functions not touched on by CPA benchmarks should also be assessed. For instance, benchmarks could check how parliament scrutinizes or examines delegated legislation. Other benchmarks could provide guidance on types of delegated legislation that are subject to affirmative resolution, which requires a vote by parliament.[11] They could also be used to ensure that delegated legislation is not overused (such overuse might be determined using statistical analyses of trends and quantitative benchmarks), that scrutiny committees are empowered to submit disallowance reports on *ultra vires* regulations for approval of the legislature, and that scrutiny committees are adequately supported by a team of legal professionals who conduct in-depth reviews of all delegated legislation on behalf of parliamentarians.

In the case of Canada, the Parliamentary Budget Office (PBO) may also provide criteria for benchmarks on parliamentary oversight functions. According to former Canadian Senator Lowell Murray (2011), over the past 40 years, the House of Commons has "allowed their most vital power, the power of the purse, to become a dead letter, their Supply and Estimates process an empty ritual." The sheer volume of estimates and the timeframes for review have made it a difficult task to carry out effectively. However, some progress was made in 2008 with the establishment of the PBO, whose mandate is to "provide independent analysis to Parliament on the state of the nation's finances, the government's estimates and trends in the Canadian economy; and upon request from a committee or parliamentarian, to estimate the financial cost of any proposal for matters over which Parliament has jurisdiction."[12] The PBO has produced several reports over the past few years to inform parliamentarians in their review of the estimates and of specific government programs. The PBO uses an integrated monitoring database that allows parliamentarians "to track the increase (or decrease) of [voted and statutory budgetary] authorities over the course of a fiscal year, as well as compare this evolution to previous years" (PBO 2011, 2).

While the above criteria touch on legislative and oversight functions, procedural functions are also an important and less understood area for which benchmarks are needed. In addition to procedural rules, there should be procedural manuals, fact sheets, scripts, and templates to assist members in their duties. As noted in the previous section, parliamentarians need records of their deliberations and agendas in the form of the daily Hansard, journals, order and notice papers, and other documentation. The Table and the Journals Office should have the expertise to provide procedural advice to the speaker, house officers, and individual members, including information regarding the daily preparation of the speaker's scroll and drafting of rulings under the speaker's direction. The members need access to lawyers with legislative drafting skills to assist them with private members' bills and amendments. They also need access to proceduralists to assist with drafting motions and navigation through the rules and daily orders of business.

Establishing Objective Benchmark Rating Scales

Rather than open-ended questions, closed (yes or no) questions, or subjective ratings, a benchmark framework should include graduated sets of criteria to reflect levels and quality of services, products, and processes. Such a rating scale should not consist of value judgments such as "strongly disagree" to "strongly agree" or "low" to "high,"[13] but rather descriptive assessment criteria against which a legislature can objectively self-assess and rank itself. The evidence of their compliance with a criteria or benchmark can be easily validated, making self-evaluation a much more objective exercise.

Although perhaps not all benchmarks lend themselves to the development of progressive criteria by which they may be rated, such criteria allow evaluators to determine how well an entity is performing on a given benchmark. To further illustrate, benchmarks on the daily production of Hansard could be set. A "yes" or "no" response would be useful in so far as the few legislatures that respond "no" will have identified a gap that must be addressed (of which they would undoubtedly be highly aware). However, criteria to establish a rating scale might set further goals for the much broader group of "yes" respondents. Some criteria might be arranged using a dashboard approach (table 11.2). The benefit of this format

Table 11.2 Example of the Dashboard Approach: Verbatim Records of Debates

Requires immediate attention	Needs improvement	Acceptable	Strong
The legislature does not have the resources in place to regularly produce verbatim transcripts of its debates.	The legislature produces verbatim transcripts of its debates but frequently incurs delays.	The legislature produces timely verbatim transcripts, which it posts and provides publicly.	In addition to timely verbatim transcripts, the legislature provides audio and video broadcasts of its debates.
The legislature produces verbatim transcripts, but they are not available to the public.	Verbatim transcripts are available to the public but are not searchable by subject matter or by parliamentarian.	Transcripts are supported by search engines that make any references to relevant subject matter publicly accessible.	Audio and video broadcasts are publicly available (including, perhaps, on demand on the legislature's website).

is that gaps are easily identified, as are any next steps that a legislature might wish to pursue on its continuous improvement agenda.

Conclusion

The benchmarking framework outlined in this chapter is just a small sample of the many sources that could help define a well-performing legislature. The standards, guidelines, benchmarks, and criteria developed by parliaments, interparliamentary assemblies, government and international development agencies, nongovernmental organizations, and academics are all useful tools that can serve multiple purposes.

Although an established benchmark framework should be complete, it should also be concise, not cumbersome. It should include objective rating criteria and cover the support services required by parliamentarians to manage and to remain fully accountable to their citizenry.

Such a framework would facilitate quality assessments that allow the setting of objectives toward strategic improvement, the identification of partnering opportunities to share services or develop capacity, and a more effective allocation of developmental funding. Because the CPA has already produced a wide range of reports assessing most aspects of parliament, it would be an excellent sponsor for such a project. The potential value of a comprehensive benchmarking tool is such that the CPA's member parliaments should rally together to develop it.

Notes

1. See comments by Agnihotri (2009, 75), which suggest that administrations are assessing the delivery of services but not the quality of those services.

2. *Small institutions* are defined in Canada as having less than a Can$300 million budget. The Senate qualifies as a small institution but has conducted internal MAF assessments annually for six years. The Treasury Board Secretariat had radically transformed the MAF for 2014.

3. This approach is similar to the European Foundation for Quality Management Excellence Model.

4. However, this report does not touch on matters of internal governance and administration.

5. The "Commonwealth (Latimer House) Principles on the Three Branches of Government" state, "An all-party committee of members of parliament should review and administer parliament's budget which should not be subject to amendment by the executive" (CPA and others 2004, 22).

6. Examples of such information include verbatim reports of debates and evidence given before committees.

7. Examples include performance reports, strategic plans, financial statements, disclosure of legislators' expenditures, and internal audit reports.

8. The information was circulated through AGORA, a leading global knowledge platform on parliamentary development. For more information about AGORA, visit http://www.agora-parl.org/.

9. See sections 6 and 7 of the Benchmarks on Legislative and Oversight Functions, which provide for certain autonomy and powers as well as conditions for effective oversight such as opposition chairs on Public Accounts Committees and reasonable time to review estimates.

10. Note, however, that much of this resource goes beyond the scope of benchmarks for parliamentary services.

11. Although an affirmative resolution procedure exists in the United Kingdom and perhaps other countries, its application in Canada is limited to user fees at present. Moreover, the mechanism in place is a negative option that allows the user fees to be deemed reported back if not studied within 20 days.

12. This quotation is from the PBO's website at http://www.pbo-dpb.gc.ca/en/ABOUT.

13. The IPU's (2008) toolkit provides for subjective ratings from "5 = very high/very good" to "1 = very low/very poor." The benefit is that when assessments are conducted by stakeholder groups (for example, parliamentarians, parliamentary staff members, the media, and the public), evaluators may be able to discern shifts of opinion on the benchmark ratings and, through further inquiry, perhaps determine the reasons behind those shifts. The pitfall is that these ratings are subjective; hence, broader samples must be used to achieve accurate assessments.

References

Agnihotri, Vivek K. 2009. "Administrative Self-Evaluation by Parliaments." Constitutional and Parliamentary Information 198, Association of Secretaries General of Parliaments, Inter-Parliamentary Union, Geneva.

———. 2010. "Quality Framework for Assessment of Parliaments: Improving Efficiency and Effectiveness." *Parliamentarian* 1: 30–35.

Brennan, Richard. 2011. "Canadians Don't Love Politicians, They Tolerate Them, Poll Shows." *Toronto Star*, March 18. http://www.thestar.com/news/canada/2011/03/18/canadians_dont_love_politicians_they_tolerate_them_poll_shows.html.

Canadian Media Research. 2006. "Report on TV Audiences for Senate Committee Hearings on CPAC." Data from the report were provided by the Cable Public Affairs Channel at an appearance before the Standing Senate Committee on Rules, Procedures, and the Rights of Parliament, November 22.

CPA (Commonwealth Parliamentary Association). 2005. "The Administration and Financing of Parliament." Report of a CPA study group hosted by the Legislature of Zanzibar, Tanzania, March 25–29, CPA and World Bank, London.

———. 2006. "Recommended Benchmarks for Democratic Legislatures." CPA, London. http://wbi.worldbank.org/wbi/Data/wbi/wbicms/files/drupal-acquia/wbi/Recommended%20Benchmarks%20for%20Democratic%20Legislatures.pdf.

———. 2010. "The CPA Benchmarks and the Parliament of Canada: A Self-Assessment." *Parliamentarian* 1 :26–29.

CPA (Commonwealth Parliamentary Association), Commonwealth Legal Education Association, Commonwealth Magistrates' and Judges' Association, and Commonwealth Lawyers' Association. 2004. "Commonwealth (Latimer House) Principles on the Three Branches of Government." Commonwealth Secretariat, London.

IPU (Inter-Parliamentary Union). 2008. "Evaluating Parliament: A Self-Assessment Toolkit for Parliaments." IPU Geneva. http://www.ipu.org/pdf/publications/self-e.pdf.

Laundy, Philip. 1989. *Parliaments in the Modern World*. London: Dartmouth.

Manning Centre. 2011. "State of Canada's Conservative Movement: September 2011." Manning Centre, Calgary. http://manningcentre.ca/wp-content/uploads/2012/11/06 -State-of-the-Movement-Report-2011.pdf.

McGee, David. 2002. *The Overseers: Public Accounts Committees and Public Spending*. London: Pluto Press.

Mendel, Toby. 2005. "Parliament and Access to Information: Working for Transparent Governance." World Bank, Washington, DC. http://siteresources.worldbank.org/WBI /Resources/Parliament_and_Access_to_Information_with_cover.pdf.

Murray, Lowell. 2011. "You Do Not Govern, You Hold to Account Those Who Do." *iPolitics*, October 13. http://www.ipolitics.ca/2011/10/13/lowell-murray-you-do-not -govern-you-hold-to-account-those-who-do/.

NDI (National Democratic Institute of International Affairs) and UNDP (United Nations Development Programme). 2004. "Strengthening Parliamentary Involvement in the Poverty Reduction Strategy Process and the Millennium Development Goals: Legislative Public Outreach on Poverty Issues." Parliaments and Poverty Toolkit 3, UNDP, New York.

PBO (Parliamentary Budget Office). 2011. "Supporting Parliamentary Scrutiny of the Estimates: The Integrated Monitoring Database (IMD). PBO, Ottawa. http://www .parl.gc.ca/PBO-DPB/documents/IMD_March_2011_EN.pdf.

CHAPTER 12

Rating the ACT Legislative Assembly against CPA Benchmarks for Democratic Legislatures: From an "A–" to an "AA"?

Wayne Berry and Tom Duncan

Introduction

In 2006, Wayne Berry, who was then speaker of the Australian Capital Territory (ACT) legislature, reviewed the Commonwealth Parliamentary Association (CPA) benchmarks, a list of 87 best practices and guidelines for self-assessments by democratic parliaments. Recognizing these standards' value for the ACT, Berry decided to conduct the jurisdiction's first CPA benchmarking exercise. The primary objective of this process was to gauge how the ACT Legislative Assembly's performance measured up against these standards and to identify areas where the ACT's form of governance could be improved. The assembly met 80 out of 87 standards, which is essentially a grade of "A–."

In 2011, Tom Duncan, clerk of the ACT Legislative Assembly, replicated the benchmarking exercise to determine whether the ACT had achieved progress in the seven shortfall areas identified by the first exercise. Indeed, this follow-up assessment suggests that the ACT has made several noteworthy enhancements and improvements between 2006 and 2011, and now deserves a grade of "A."

This chapter aims to share the main findings from the ACT Legislative Assembly's first CPA benchmarking exercise in 2006 and the follow-up assessment in 2011. The chapter is organized as follows: The first section discusses the ACT's first experience conducting the CPA benchmarking exercise. The second section assesses progress made since the first exercise. The final section concludes by discussing how the benchmarking exercise has been useful for the ACT.

The authors would like to acknowledge the work of David Skinner, manager of strategy and parliamentary education, Australian Capital Territory Legislative Assembly Secretariat.

The First Benchmark Exercise: 2006

The 2006 benchmarking exercise employed a broad interpretation of each benchmark, and considered compliance with both the letter and spirit of the measures set out by the CPA. See annex 12A for a complete list of the relevant benchmarks and the assembly's performance against them.

Out of the 87 benchmarks, seven underperforming areas were identified (a grade of A–): (a) assembly budget, (b) legislative debate, (c) committee review, (d) independent employment arrangements, (e) staff code of conduct, (f) committee oversight, and (g) public votes. This section reviews the challenges faced in these areas.

Assembly Budget

The first issue that the 2006 benchmarking exercise identified was the lack of autonomy in formulating the Legislative Assembly's budget. CPA benchmark 6.1.2 states, "Only the legislature shall be empowered to determine and approve the budget of the legislature." The Latimer House Principles provide similar guidelines concerning the development and administration of parliamentary budgets: "An all-party committee of members of parliament should review and administer parliament's budget which should not be subject to amendment by the executive" (CPA and others 2004, 22).

Despite these best-practice recommendations, Australia's executive was heavily involved in the Legislative Assembly's budget process at the time of the first self-assessment. The assembly's Standing Committee on Administration and Procedure helped develop the assembly's budget submission, but the executive, through the Budget Cabinet, unilaterally decided the amount of funding to be inserted in the appropriation bill. Although the assembly could then vote on the appropriation bill and recommend amendments, when there was a majority government, an appropriation bill would almost always be passed in its original form without amendment.

The executive branch's incursion into the legislative branch's affairs was thus a critical area for reform, because it fundamentally affected the proper expression of the separation of powers doctrine.

Legislative Debate

The second area for improvement relates to CPA benchmark 2.5.2: "The legislature shall provide adequate opportunity for legislators to debate bills prior to a vote." The Legislative Assembly largely conformed to this measure; however, its right to make a closure or "gag" motion compromised the adequacy of debate on legislation. As in many other parliaments, a majority government[1] could end a debate and resolve a question immediately by applying a closure motion to a particular item under discussion, including a bill.

Closure motions were not commonly used in the assembly. From 1996 to 2006, only three bills were declared urgent, and from 2000 to 2006, no bills were subject to a closure motion. Nonetheless, because parliamentary procedures did allow for the opportunity for debate to be compromised, the assembly did not meet this benchmark.

Committee Review

The third area of concern relates to CPA benchmark 3.2.1: "There shall be a presumption that the legislature will refer legislation to a committee, and any exceptions must be transparent, narrowly defined, and extraordinary in nature." The Legislative Assembly did not comply with this measure, if it is taken to mean that all bills are referred for substantive review on the policy aspects of the legislation by a standing, select, or committee-as-whole type apparatus.

The assembly did, however, refer all legislation to its Standing Committee on Scrutiny of Bills and Subordinate Legislation. Among other roles, this committee evaluated and reported on whether legislation unduly trespassed on personal rights and liberties or inappropriately delegated legislative powers and insufficiently subjected the exercise of legislative power to parliamentary scrutiny.

It was not the Standing Committee on Scrutiny of Bills and Subordinate Legislation's responsibility to form a view on the merits of the public policy dimensions expressed in the legislation. To refer each piece of legislation for substantive review along these lines could result in legislative gridlock and did not seem desirable to any assembly members.

Independent Employment Arrangements

The fourth area for improvement relates to CPA benchmark 5.1.2: "The legislature, rather than the executive branch, shall control the parliamentary service and determine the terms of employment." The ACT did not conform to this benchmark for a number of reasons. In particular, secretariat staff comprised ACT government (executive) personnel employed under the Public Sector Management Act 1994. Moreover, these staff members were bound by the ACT Public Service Code of Conduct,[2] and their rates of pay and conditions flowed from template agreements negotiated at the whole-of-government level by the executive.

However, the independence of the clerk (the secretariat's administrative head) and his or her staff was protected through several other provisions in both the Public Sector Management Act and the Financial Management Act 1996. Furthermore, at the time of the 2006 benchmarking exercise, it was an open question as to whether a stand-alone legislative framework should be devised for the secretariat staff. Although separating the functions performed by secretariat staff from those of the wider civil service had advantages, considerable administrative overhead would be involved in developing a further set of industrial and governance-related policies.

Staff Code of Conduct

The fifth area for improvement relates to CPA benchmark 5.4.3: "All staff shall be subject to a code of conduct." In the assembly's case, secretariat staff and assembly members are both subject to a code of conduct. However, the assembly had not implemented a code of conduct for the members' staff.

A code of conduct for the members' staff is needed to outline general principles and standards of behavior. Indeed, the assembly's Standing Committee on Administration and Procedure recommended that such a code be developed a number of years ago. Although political sanctions can be levied against assembly members and any staff members who do not observe general community standards of behavior, a specific code for the staff (endorsed by members and given continuing effect by the assembly) would provide an explicit covenant that more legitimately binds assembly members' offices to proper standards.

Committee Oversight

The sixth area of concern relates to CPA benchmark 7.2.2: "Oversight committees shall provide meaningful opportunities for minority or opposition parties to engage in effective oversight of government expenditures. Typically, the public accounts committee will be chaired by a member of the opposition party." The rationale behind this benchmark is that the government's expenditure and revenue decisions are subject to rigorous scrutiny when an opposition member heads the public accounts committee (PAC).

This convention had been consistently observed until 2007, when the opposition chair was deposed by a vote of the committee (made up of three members—one opposition member, one government member, and one member of the crossbench). This change was the legitimate prerogative of the committee.

However, the assembly still lived up to the benchmark, at least its spirit, with this change, because the opposition chair was replaced by a crossbench chair (a member of the Green Party), rather than a government chair. Since self-government in the ACT, there have been 12 PAC chairs. The first chair was a member of the government, whereas the chair at the time of the 2006 assessment was a crossbench member. In between, however, 10 chairs had been members of the opposition.[3]

It is also worth noting that the former majority government (prior to the first benchmarking exercise) had observed the convention that the deputy speaker be an opposition member, despite having sufficient numbers to award both the speakership and deputy speakership to government members. Eschewing a "winner takes all" approach is inherently democratic, and conventions such as these form important benchmarks with which to assess the democratic character of a legislature.

Public Votes

The final area of concern that emerged from the first benchmarking exercise relates to CPA benchmark 2.6.1: "Plenary votes in the legislature shall be public." At the time of the first exercise, the Legislative Assembly generally observed this benchmark but also maintained a number of exceptions. In particular, when a new assembly commenced, the election of the speaker, deputy speaker, and chief minister was conducted by secret ballot; thus, the voting records of individual members could not be publicly known.

Arguments can be made for and against secret ballots. One advantage of secret ballots is that they alleviate any external pressure on a member to vote for a particular candidate. Conversely, an inherent part of a pluralist, democratic process is that constituents and interest groups can persuade elected representatives to cast their votes in particular ways. Indeed, many argue that the public has a right to know how assembly members vote in all aspects of their public duty and that making ballots secret impairs the accountability of members to their constituents. For example, a member could publicly support a politically popular candidate for speaker but, for whatever reason, hide behind the secret ballot to vote for a different candidate. This lack of transparency has the potential to thwart the assembly's accountability.

The Second Benchmark Exercise: 2011

In 2011, a second exercise assessed whether any progress had been made to address the deficiencies identified in the first CPA self-assessment. This section reviews developments made in the seven areas identified as shortfalls in 2006.

Assembly Budget Revisited

In the first benchmarking exercise, Speaker Berry expressed great concern about the lack of autonomy in the legislature's budget development and decision-making process. According to CPA benchmark 6.1.2, the legislature needed greater budget control to improve the assembly's democratic credentials and give more fulsome expression to the separation of powers doctrine.

Between 2006 and 2011, no formal or legislative changes were made to the way that the legislature secures funding, but the culture and practices for developing the assembly budget did change. For example, both the executive and the Department of the Treasury now recognize that the legislature should not be subject to the arbitrary will of the executive when its budget is being formulated. Therefore, since 2009 at least, the assembly's budget has not been reduced. Indeed, when the speaker requested additional funding for the legislature from the treasurer in the 2009/10 financial year, the funding was included in the budget. Furthermore, when the executive decided that, because of the global financial crisis, it needed to impose an efficiency dividend on all government agencies, the treasurer wrote to the speaker to ask whether the assembly would be willing to participate in such a cost-cutting program. When the speaker replied that the assembly would not participate but instead would identify savings itself, no objection was raised.

So although a formal, binding process to meet this benchmark has yet to be implemented, some progress has been made, and it appears that the executive is observing the benchmark's spirit in its dealings with the legislature over the budget. In other words, the underlying principle behind the benchmark has been recognized through evolving informal conventions.

Moreover, the speaker plans to table legislation that would provide an independent staffing structure for the Legislative Assembly and possibly codify separate budget arrangements for the assembly.[4] Although a move in this direction would help the assembly meet this benchmark, until the move occurs the benchmark cannot be considered to have been fully met.

Legislative Debate Revisited

In the first benchmarking exercise, Speaker Berry observed that the right to apply a closure motion conflicted with CPA benchmark 2.5.2, which highlights the need for adequate opportunity to debate bills prior to a vote. The right to make a closure motion leaves neither opportunity for members to consider the legislation nor sufficient time to consult affected organizations or the community at large.

In December 2008, the assembly agreed on a temporary order to ensure that a bill in principle stage (also known in some legislatures as the *second reading*) could not be agreed to in the same sitting period in which it was introduced. In effect, this order means that when a bill is introduced into the assembly, it cannot be debated again for at least three weeks. Although three weeks is still a quick turnaround relative to practices in some legislatures, it does mark some degree of progress against the CPA benchmark. Moreover, if the temporary order is ultimately adopted as a standing order, it will ensure that a process exists to provide an adequate opportunity for legislators to examine and consult on bills prior to debating and voting on them.

It should also be noted that only the annual appropriation bill (the main budget bill of the year) has been declared "urgent" (that is, a guillotine or gag motion is put in place) since 2007. However, even though the declaration of urgency motion passed, members still had almost 15 hours of debate on the bill (together with 107 hours spent in Estimates Committee considering the bill), which most members would view as adequate opportunity to debate a bill prior to a vote. Between 2006 and 2011, all bills considered by the assembly have undergone debate unconstrained by urgency motions.

Committee Review Revisited

The assembly made little progress on the third shortfall identified in the first exercise (the one related to CPA benchmark 3.2.1 on the need for committee review of legislation). Since the commencement of the Seventh Assembly in December 2008, only 7 bills out of a total of 173 (4 percent) have been referred to a committee.

Independent Employment Arrangements Revisited

The assembly has achieved noteworthy progress against CPA benchmark 5.1.2, which proposes that parliamentary employment be controlled by the legislature as opposed to the executive. In February 2011, the speaker notified all assembly members of plans to introduce legislation that would establish independence from the executive of staff members working for the Legislative Assembly Secretariat (the parliamentary service). This bill cements control of the secretariat staff as a responsibility of the clerk of the assembly and clearly differentiates staff in service of the legislature from those in service of the executive government. The speaker's announcement received positive feedback and signs that the passage of legislation in this area is feasible. Based on these developments, it is anticipated that the assembly will meet this benchmark in the near future.

Staff Code of Conduct Revisited
Little progress has been made against CPA benchmark 5.4.3, which advocates for a staff code of conduct. Although assembly members and secretariat staff are subject to a code of conduct, the staff does not have a publicly available code. Instead, the employment contracts of staff personnel have a clause that sets out a code of conduct that must be observed.

Committee Oversight Revisited
In the first self-assessment, the assembly did not meet CPA benchmark 7.2.2 on the need for minority or opposition parties to oversee government expenditures. In 2006, the PAC chair was not a member of the opposition party, but instead a member of a minor party occupying a position on the crossbenches of the assembly. Moreover, government members had chaired the Select Committee on Estimates for the previous four years.

In the current assembly, the chair is a minor party member. Of the six general-purpose committees, three are chaired by crossbench members, two are chaired by opposition members, and one is chaired by a government member. In other words, 83 percent of committees are chaired by nongovernment members.

It should also be noted that the assembly forms a select committee each year to examine the appropriation bill. The three committees formed between 2008 and 2011 have been chaired by the leader of the opposition, the deputy leader of the opposition, and a crossbench member, respectively.

This background indicates that the chairing arrangements in place within the oversight committees provide abundant opportunities for executive oversight and scrutiny of government expenditure. Indeed, the fact that the PAC has had such a long history of being chaired by nongovernment members (predominantly opposition members, but also crossbench members) fulfills the spirit of this benchmark.

Public Votes Revisited
The final area for improvement identified in the first benchmarking exercise relates to CPA benchmark 2.6.1, which proposes that plenary votes shall be public. Speaker Berry found that the assembly generally observed the requirement for votes being public, but that the assembly should be marked down for

not organizing public votes for the election of officer bearers such as the chief minister, the speaker, or the deputy speaker.

Since the 2006 exercise, a relevant endnote to benchmark 2.6.1 was discovered. The endnote states, "The Study Group noted that one possible exception to this may be the election of office bearers." Hence, the assembly in fact meets this particular benchmark, because all votes in the plenary legislature, apart from the election of these office bearers, are public votes.

Conclusion

This chapter aimed to share the main findings from the ACT Legislative Assembly's first CPA benchmarking exercise in 2006 and follow-up assessment in 2011. In the first exercise, the ACT obtained satisfactory results against 80 of the 87 benchmarks contained in the CPA list. The remaining seven benchmarks were listed as either not fully meeting the spirit or the letter of the measure set out by the CPA study group. According to Speaker Berry, the main area requiring urgent attention was the legislature's budget development and decision-making process.

The second review exercise revealed a noteworthy improvement in the assembly's rating, from 80 out of 87 in 2006 to 84 out of 87 in 2011. Assuming that the 2006 tally resulted in a grade of "A–" for the ACT Legislative Assembly, a rating of an "A" is justified for 2011 in view of the enhancements and improvements made over the years between the two assessments.

Was the benchmarking exercise useful? At the time of the original assessment in 2006, the ACT was the first jurisdiction to attempt the CPA benchmarking exercise. The assessment was a very useful tool for the legislature, because it helped identify areas in which the legislature complies with the benchmarks as well as areas where further attention is needed. As outlined earlier in this chapter, progress has been made against some of the benchmarks that were assessed as shortfalls in 2006, and tracking such progress can only enhance the governance of the jurisdiction.

The benchmarking exercise has also promoted awareness about democratic benchmarks and assisted in shaping elements of the institutional and political culture, insofar as the separation of powers is concerned. As a result, politicians and senior public servants have a greater understanding of the importance of the doctrine and its application to the ACT's form of government.

Annex 12A: Results of the First Benchmarking Exercise

The first CPA benchmarking exercise in the ACT was held in December 2006. Table 12A.1 lists the benchmarks against which the legislature was assessed and shows the results of that assessment.

Table 12A.1 Benchmarks for Democratic Legislatures: Compliance Results for the Australian Capital Territory

Benchmark	Was the legislature found compliant?	Comments
1.1.1 Members of the popularly elected or only house shall be elected by direct universal and equal suffrage in a free and secret ballot.	Yes	
1.1.2 Legislative elections shall meet international standards for genuine and transparent elections.	Yes	
1.1.3 Term lengths for members of the popular house shall reflect the need for accountability through regular and periodic legislative elections.	Yes	Fixed-term elections are held every four years.
1.2.1 Restrictions on candidate eligibility shall not be based on religion, gender, ethnicity, race, or disability.	Yes	
1.2.2 Special measures to encourage the political participation of marginalized groups shall be narrowly drawn to accomplish precisely defined, and time-limited, objectives.	Yes	No special measures of this nature are in place.
1.3.1 No elected member shall be required to take a religious oath against his or her conscience in order to take his or her seat in the legislature.	Yes	Members can take an oath or a secular affirmation.
1.3.2 In a bicameral legislature, a legislator may not be a member of both houses.	n.a.	The assembly is unicameral.
1.3.3 A legislator may not simultaneously serve in the judicial branch or as a civil servant of the executive branch.	Yes	
1.4.1 Legislators shall have immunity for anything said in the course of the proceedings of legislature.	Yes	Parliamentary privilege applies to assembly members.
1.4.2 Parliamentary immunity shall not extend beyond the term of office, but a former legislator shall continue to enjoy protection for his or her term of office.	Yes	
1.4.3 The executive branch shall have no right or power to lift the immunity of a legislator.	Yes	
1.4.4 Legislators must be able to carry out their legislative and constitutional functions in accordance with the constitution, free from interference.	Yes	
1.5.1 The legislature shall provide proper remuneration and reimbursement of parliamentary expenses to legislators for their service, and all forms of compensation shall be allocated on a nonpartisan basis.	Yes	

table continues next page

Benchmarking and Self-Assessment for Parliaments • http://dx.doi.org/10.1596/978-1-4648-0327-7

Table 12A.1   Benchmarks for Democratic Legislatures: Compliance Results for the Australian Capital Territory (continued)

Benchmark	Was the legislature found compliant?	Comments
1.6.1 Legislators shall have the right to resign their seats.	Yes	
1.7.1 The legislature shall have adequate physical infrastructure to enable members and staff to fulfill their responsibilities.	Yes	This area is difficult to assess and relates to the Latimer House principles. However, in general, the assembly has adequate physical infrastructure (that is, there is a building and associated facilities, which are in good repair and provide a suitable venue for the assembly and its committees to undertake their work effectively).
2.1.1 Only the legislature may adopt and amend its rules of procedure.	Yes	
2.2.1 The legislature shall select or elect presiding officers pursuant to criteria and procedures clearly defined in the rules of procedure.	Yes	
2.3.1 The legislature shall meet regularly, at intervals sufficient to fulfill its responsibilities.	Yes	
2.3.2 The legislature shall have procedures for calling itself into regular session.	Yes	
2.3.3 The legislature shall have procedures for calling itself into extraordinary or special session.	Yes	
2.3.4 Provisions for the executive branch to convene a special session of the legislature shall be clearly specified.	Yes	Nine members of the assembly must agree to a special sitting of the assembly proceeding (that is, an absolute majority).
2.4.1 Legislators shall have the right to vote to amend the proposed agenda for debate.	Yes	
2.4.2 Legislators in the lower or only house shall have the right to initiate legislation and to offer amendments to proposed legislation.	Yes	
2.4.3 The legislature shall give legislators adequate advance notice of session meetings and the agenda for the meeting.	Yes	
2.5.1 The legislature shall establish and follow clear procedures for structuring debate and determining the order of precedence of motions tabled by members.	Yes	

table continues next page

Table 12A.1 Benchmarks for Democratic Legislatures: Compliance Results for the Australian Capital Territory *(continued)*

Benchmark	Was the legislature found compliant?	Comments
2.5.2 The legislature shall provide adequate opportunity for legislators to debate bills prior to a vote.	No	As is the case in many democratic legislatures, adequate opportunity for debate on bills can be curtailed by the application of a closure motion or "gag" by the majority party.
2.6.1 Plenary votes in the legislature shall be public.	No	The only exception to this benchmark is the election of the speaker, deputy speaker, and chief minister at the commencement of a new assembly. The election of these officers, while still a public proceeding, is conducted by secret ballot.
2.6.2 Members in a minority on a vote shall be able to demand a recorded vote.	Yes	
2.6.3 Only legislators may vote on issues before the legislature.	Yes	
2.7.1 The legislature shall maintain and publish readily accessible records of its proceedings.	Yes	
3.1.1 The legislature shall have the right to form permanent and temporary committees.	Yes	
3.1.2 The legislature's assignment of committee members on each committee shall include both majority and minority party members and reflect the political composition of the legislature.	Yes	
3.1.3 The legislature shall establish and follow a transparent method for selecting or electing the chairs of committees.	Yes	
3.1.4 Committee hearings shall be in public. Any exceptions shall be clearly defined and provided for in the rules of procedure.	Yes	
3.1.5 Votes of committee shall be in public. Any exceptions shall be clearly defined and provided for in the rules of procedure.	Yes	
3.2.1 There shall be a presumption that the legislature will refer legislation to a committee, and any exceptions must be transparent, narrowly defined, and extraordinary in nature.	No	This is not the case in the assembly.
3.2.2 Committees shall scrutinize legislation referred to them and have the power to recommend amendments or amend the legislation.	Yes	

table continues next page

Table 12A.1 Benchmarks for Democratic Legislatures: Compliance Results for the Australian Capital Territory (continued)

Benchmark	Was the legislature found compliant?	Comments
3.2.3 Committees shall have the right to consult and/or employ experts.	Yes	
3.2.4 Committees shall have the power to summon persons, papers, and records, and this power shall extend to witnesses and evidence from the executive branch, including officials.	Yes	
3.2.5 Only legislators appointed to the committee, or authorized substitutes, shall have the right to vote in committee.	Yes	
3.2.6 Legislation shall protect informants and witnesses presenting relevant information to commissions of inquiry about corruption or unlawful activity.	Yes	
4.1.1 The right of freedom of association shall exist for legislators, as for all people.	Yes	
4.1.2 Any restrictions on the legality of political parties shall be narrowly drawn in law and shall be consistent with the International Covenant on Civil and Political Rights.	Yes	
4.2.1 Criteria for the formation of parliamentary party groups, and their rights and responsibilities in the legislature, shall be clearly stated in the rules.	Yes	
4.2.2 The legislature shall provide adequate resources and facilities for party groups pursuant to a clear and transparent formula that does not unduly advantage the majority party.	Yes	
4.3.1 Legislators shall have the right to form interest caucuses around issues of common concern.	Yes	Although not prescribed, interest caucuses are not prohibited.
5.1.1 The legislature shall have an adequate nonpartisan professional staff to support its operations, including the operations of its committees.	Yes	
5.1.2 The legislature, rather than the executive branch, shall control the parliamentary service and determine the terms of employment.	No	This is not the case in many respects. Although recognized as being independent of executive government, secretariat staff members are essentially Australian Capital Territory (ACT) government public servants with the same terms of employment, which derive from agreements made with the government of the day.

table continues next page

Table 12A.1 Benchmarks for Democratic Legislatures: Compliance Results for the Australian Capital Territory (continued)

Benchmark	Was the legislature found compliant?	Comments
5.1.3 The legislature shall draw and maintain a clear distinction between partisan and nonpartisan staff.	Yes	
5.1.4 Members and staff of the legislature shall have access to sufficient research, library, and ICT [information, communication, and technology] facilities.	Yes	It is difficult to define *sufficient*. In whose mind? There could well be individual members who view the standard of facilities as being insufficient. However, in general, assembly members and staff members have access to these facilities to a reasonable standard, and concerns about the sufficiency of facilities can be addressed through the assembly's Standing Committee on Administration and Procedure, which advises the speaker on members' entitlements and facilities.
5.2.1 The legislature shall have adequate resources to recruit staff sufficient to fulfill its responsibilities. The rates of pay shall be broadly comparable to those in the public service.	Yes	
5.2.2 The legislature shall not discriminate in its recruitment of staff on the basis of race, ethnicity, religion, gender, disability, or, in the case of nonpartisan staff, party affiliation.	Yes	
5.3.1 Recruitment and promotion of nonpartisan staff shall be on the basis of merit and equal opportunity.	Yes	
5.4.1 The head of the parliamentary service shall have a form of protected status to prevent undue political pressure.	Yes	
5.4.2 Legislatures should, either by legislation or resolution, establish corporate bodies responsible for providing services and funding entitlements for parliamentary purposes and providing for governance of the parliamentary service.	Yes	The ACT Legislative Assembly Secretariat is recognized in the Public Sector Management Act and the Financial Management Act as having an independent status in supporting the work of the legislature.
5.4.3 All staff shall be subject to a code of conduct.	No	The assembly partly complies. All secretariat staff members are subject to both the ACT Public Service Code of Conduct and a secretariat-specific code of conduct. However, staff personnel employed by assembly members are not subject to an assembly-specific code of conduct.

table continues next page

Table 12A.1 Benchmarks for Democratic Legislatures: Compliance Results for the Australian Capital Territory *(continued)*

Benchmark	Was the legislature found compliant?	Comments
6.1.1 The approval of the legislature is required for the passage of all legislation, including budgets.	Yes	
6.1.2 Only the legislature shall be empowered to determine and approve the budget of the legislature.	No	The assembly does not comply with the spirit of this measure. In practice, the government of the day determines the quantum of funding made available through the appropriation bill.
6.1.3 The legislature shall have the power to enact resolutions or other nonbinding expressions of its will.	Yes	
6.1.4 In bicameral systems, only a popularly elected house shall have the power to bring down government.	n.a.	
6.1.5 A chamber where a majority of members are not directly or indirectly elected may not indefinitely deny or reject a money bill.	n.a.	
6.2.1 In a bicameral legislature, there shall be clearly defined roles for each chamber in the passage of legislation.	n.a.	
6.2.2 The legislature shall have the right to override an executive veto.	n.a.	
6.3.1 Opportunities shall be given for public input into the legislative process.	Yes	
6.3.2 Information shall be provided to the public in a timely manner regarding matters under consideration by the legislature.	Yes	
7.1.1 The legislature shall have mechanisms to obtain information from the executive branch sufficient to exercise its oversight function in a meaningful way.	Yes	
7.1.2 The oversight authority of the legislature shall include meaningful oversight of the military security and intelligence services.	n.a.	
7.1.3 The oversight authority of the legislature shall include meaningful oversight of state-owned enterprises.	Yes	However, claims of commercial-in-confidence status have been viewed as an impediment to oversight throughout the years.
7.2.1 The legislature shall have a reasonable period of time in which to review the proposed national budget.	Yes	

table continues next page

Table 12A.1 Benchmarks for Democratic Legislatures: Compliance Results for the Australian Capital Territory (continued)

Benchmark	Was the legislature found compliant?	Comments
7.2.2 Oversight committees shall provide meaningful opportunities for minority or opposition parties to engage in effective oversight of government expenditures. Typically, the public accounts committee will be chaired by a member of the opposition party.	No	The assembly has a proud record with respect to this benchmark. Although in many other jurisdictions governments have used their numbers to install a government chair to this position, the assembly has consistently had an opposition chair in the role. However, recently the opposition chair of the Public Accounts Committee was deposed by a vote of its membership and a member of the crossbench assumed the chair. One can argue that the assembly is meeting the spirit of the benchmark because the chair remains a nongovernment member.
7.2.3 Oversight committees shall have access to records of executive branch accounts and related documentation sufficient to be able to meaningfully review the accuracy of executive branch reporting on its revenues and expenditures.	Yes	
7.2.4 There shall be an independent, nonpartisan supreme or national audit office whose reports are tabled in the legislature in a timely manner.	Yes	
7.2.5 The supreme or national audit office shall be provided with adequate resources and legal authority to conduct audits in a timely manner.	Yes	
7.3.1 The legislature shall have mechanisms to impeach or censure officials of the executive branch or express no confidence in the government.	Yes	
7.3.2 If the legislature expresses no confidence in the government, the government is obliged to offer its resignation. If the head of state agrees that no other alternative government can be formed, a general election should be held.	Yes	
8.1.1 The legislature shall provide all legislators with adequate and appropriate resources to enable the legislators to fulfill their constituency responsibilities.	Yes	

table continues next page

Table 12A.1 **Benchmarks for Democratic Legislatures: Compliance Results for the Australian Capital Territory** *(continued)*

Benchmark	Was the legislature found compliant?	Comments
8.2.1 The legislature shall have the right to receive development assistance to strengthen the institution of parliament.	n.a.	
8.2.2 Members and staff of parliament shall have the right to receive technical and advisory assistance, as well as to network and exchange experience with individuals from other legislatures.	Yes	
9.1.1 The legislature shall be accessible and open to citizens and the media, subject only to demonstrable public safety and work requirements.	Yes	
9.1.2 The legislature should ensure that the media are given appropriate access to the proceedings of the legislature without compromising the proper functioning of the legislature and its rules of procedure.	Yes	
9.1.3 The legislature shall have a nonpartisan media relations facility.	Yes	The clerk is responsible for providing general nonpartisan information about the assembly when media requests are made. The speaker of the assembly makes media comments about the specific operations of the legislature, again, in a nonpartisan way.
9.1.4 The legislature shall promote the public's understanding of the work of the legislature.	Yes	
9.2.1 Where the constitution or parliamentary rules provide for the use of multiple working languages, the legislature shall make every reasonable effort to provide for simultaneous interpretation of debates and translation of records.	n.a.	
10.1.1 Legislators should maintain high standards of accountability, transparency, and responsibility in the conduct of all public and parliamentary matters.	Yes	This is a matter of community perception and debate and is not easily assessed. However, the assembly has recently established an ethics adviser position, which will provide a source of advice and information on areas of ethical ambiguity and will add extra assurance that assembly members uphold high standards of accountability, transparency, and responsibility.

table continues next page

Table 12A.1 Benchmarks for Democratic Legislatures: Compliance Results for the Australian Capital Territory *(continued)*

Benchmark	Was the legislature found compliant?	Comments
10.1.2 The legislature shall approve and enforce a code of conduct, including rules on conflicts of interest and the acceptance of gifts.	Yes	There is a code of conduct for assembly members. There are also rules and procedures for declaring gifts.
10.1.3 Legislatures shall require legislators to fully and publicly disclose their financial assets and business interests.	Yes	Members are required to declare any gifts or other financial or business interests through a Statement of Registrable Interests form. Completed forms are kept by the clerk of the assembly and are accessible to the public and the press on request. The purpose of the form is to place on the public record members' and ministers' interests that may conflict, or may be seen to conflict, with their public duty.
10.1.4 There shall be mechanisms to prevent, detect, and bring to justice legislators and staff engaged in corrupt practices.	Yes	Although the ACT has no independent commission to investigate corruption, the Australian Federal Police has a remit to review suspected breaches of the criminal law perpetrated by assembly members. The ACT auditor general and ombudsman also play a role in receiving and investigating reports concerning assembly members suspected of wrongdoing.

Notes

1. In the Legislative Assembly's case, a majority government holds 9 or more of the 17 seats.

2. Staff members were also bound by a secretariat-specific code of conduct.

3. Nine of these chairs were either leaders of the opposition or shadow treasurers.

4. As discussed later in this section, there is a possibility that this effort will include a separate budget appropriation for the assembly.

Reference

CPA (Commonwealth Parliamentary Association), Commonwealth Legal Education Association, Commonwealth Magistrates' and Judges' Association, and Commonwealth Lawyers' Association. 2004. "Commonwealth (Latimer House) Principles on the Three Branches of Government." Commonwealth Secretariat, London.

CHAPTER 13

Assessing Parliament Using the CPA Benchmarks and the IPU Toolkit: A Personal Perspective from Kiribati

Hon. Taomati Iuta

Background

I was first elected to the Parliament of Kiribati in 1978. In 2008, I was approached by the Commonwealth Parliamentary Association (CPA), which asked me to assess my parliament using the CPA benchmarks framework. Truthfully, my immediate reaction was one of anger, because to me the word *benchmark* meant a measure of excellence. In my mind, asking me to undergo a benchmarking exercise was an indirect way of suggesting that my parliament was not up to some accepted standard.

Rightly or wrongly, I have always thought that my parliament conducted its affairs in accordance with democratic principles, and therefore I was not prepared to compare how it performed against the standard of some other parliament that was selected as a "shining example." After the CPA's suggestion, my mind went into overdrive trying to determine what other parliaments had done wrong, and I fortified my conviction that Kiribati's parliament was better than all other parliaments. Concluding that there was no point in undertaking a benchmarking exercise, I prepared myself to forget about it.

Soon thereafter, however, the CPA secretary general suggested organizing a workshop on benchmarking to accompany the "First among Equals" workshop at the Queensland Parliament in Brisbane in June 2009, which was a professional development course for speakers organized by the Centre for Democratic Institutions. The benchmarking workshop was to be conducted by Andrew Imlach of the CPA Secretariat in London and by Alifereti Bulivou of the Pacific office of the Pacific Parliamentarians Association on Population and Development and the Forum Presiding Officers Conference office in Suva.

This chapter, which recounts how I learned about the CPA benchmarking exercise at this workshop and applied it to the case of my parliament in Kiribati, is organized as follows: The first section discusses my experience participating in the 2009 Brisbane workshop and learning about the benchmarking exercise's objective. The following section focuses on the benchmarking process in Kiribati. The subsequent section highlights several key advancements that emerged from the benchmarking exercise in Kiribati, as well as one key area where further improvement is needed. The concluding section discusses how the benchmarking assessment is aligned with the vision of the Parliament of Kiribati.

Objective of the CPA Benchmarking Exercise

At the Brisbane benchmarking workshop in 2009, the organizers first explained how the benchmark concept has evolved in the parliamentary context. Contrary to my perception, they noted that the benchmarking exercise does not classify any parliament as an ideal, nor does it require other parliaments to compare themselves to this ideal. Rather, this exercise is designed to ask pertinent questions that have been thoughtfully compiled to support and encourage the good operation of a democratically sound institution.

The organizers explained that there are two main reasons behind the formulation of the CPA benchmarks. First, when undertaking the benchmarking exercise, one registers a keen interest in adhering to democratic parliamentary principles. In so doing, one signals this interest to world bodies and aid donors that assist in the development of democratic parliaments. Such assistance can strengthen the capacity of parliaments to play their role more effectively in representing the people and in scrutinizing the performance of the executive. Second, benchmarking reinforces self-assessment. One is not comparing one's own parliament to another parliament but instead is undertaking an internal assessment exercise. As in everyday life, it is a healthy activity to pursue self-improvement by setting personal targets or benchmarks.

After explaining the CPA benchmarking exercise, the organizers divided the workshop participants into three smaller groups and asked them to respond to their selected areas of benchmarking. The resource personnel were also divided into three groups and assisted the participant groups.

Each group came up with answers regarding how their parliaments had performed. If the group could not reach a consensus, a general discussion was opened up to find a common answer that all members supported. To conclude this session, the groups presented their answers back to plenary, with one member of each group acting as a spokesperson. All participants were allowed to make comments on each of the group presentations.

The exercise helped all members of the workshop to fully understand the aim of benchmarking. In the end, the participants reached a general consensus that benchmarking was a valuable exercise. We agreed that all Commonwealth Pacific parliaments should attempt to complete the CPA benchmarking exercise before the end of 2009.

Benchmarking and Self-Assessment for Parliaments • http://dx.doi.org/10.1596/978-1-4648-0327-7

Practical Applications Leading to Valuable Assessments

I am happy to report that Kiribati was one of the first of the small Pacific island parliaments to complete the benchmarking exercise. This effort was possible with the assistance of Alifereti Bulivou, who, before coming to Brisbane, had contacted our parliament suggesting that he assist us in our benchmarking self-assessment during his upcoming visit to Kiribati for the CPA Pacific Region Presiding Officers and Clerks Conference. As I was leaving for Brisbane, I told the clerk to inform Mr. Bulivou that his benchmarking idea would not be welcome. However, after participating in the benchmarking exercise in Brisbane, I completely changed my mind and was convinced that it would be a worthwhile exercise for my parliament. I therefore agreed with Mr. Bulivou to undertake this exercise with my parliament on July 3, 2009, the day after he arrived in Tarawa.

For Kiribati's first benchmarking exercise, I invited parliamentarians and several public servants to participate. Although there was insufficient time to get as many people as I would have liked, an adequate level of parliamentary involvement was still achieved. On the morning of Friday, July 3, Mr. Bulivou and I met with all members of the parliamentary senior staff, who concurred that benchmarking should be done. That afternoon, we met again and were joined by a few of the parliamentarians who were present in Tarawa (including the secretary to the Cabinet and the attorney general). During this afternoon session, we covered all of the benchmarking questions, and we agreed to reconvene the following Monday to continue our discussion with Mr. Bulivou.

On Monday, we systematically answered the benchmark questions. However, we realized that our views were not necessarily representative of the views of the parliamentarians who had been unable to attend. As a result, we agreed that Mr. Bulivou would take notes during our discussions and that we would distribute these notes when all the remaining parliamentarians arrived for the next parliamentary meeting. Only after the other parliamentarians had provided their comments would the benchmarking exercise be complete.

At the next sitting of parliament, Mr. Bulivou's benchmarking notes were tabled at a special general meeting of all the parliamentarians, with particular attention given to members who had not been present at the benchmarking exercise. On the whole, the benchmarking of the Parliament of Kiribati was endorsed as presented (see annex 13A).[1] Only a few outstanding issues needed further discussion at a later, more appropriate time.

A Higher Standard for Kiribati's Parliament

The CPA benchmarking exercise has encouraged the Parliament of Kiribati to make its parliamentarians fully understand their role as legislators, scrutinizers, and representatives for their constituencies. This section (a) highlights two advancements in parliamentary procedure that have emerged since the benchmarking exercise was undertaken and (b) discusses one area in which further improvement is still needed.

A First Advancement: Greater Scrutiny of the Executive

One of the most important tasks of parliamentarians is to scrutinize the work of the executive so that corrupt practices are detected in their early stages and any mismanagement of public funds can be probed carefully and corrected in a timely manner. Triggered, at least in part, by shortfalls in this scrutiny process, the United Nations Development Programme recommended that the Kiribati government review the rules of parliament procedures. Soon thereafter, a parliamentary committee was established to undertake this review. Kiribati's new rules of procedure that emerged from this review have strengthened the scrutiny process by increasing question time from one hour to two hours at every sitting day except on the first day of the session. Under the new rules, the speaker may allow a member to ask a question without notice on a Wednesday sitting if it is of an urgent nature and relates to matters of public importance to the nation.

Of particular importance for the increasing accountability of the executive is the new rule 65(b) because it forces the government to take notice of the reports of the Public Accounts Committee (PAC) and to act on the PAC's recommendations immediately after the reports are published. Moreover, members of parliament are expected to scrutinize the government's response and to raise pertinent questions or motions. This opportunity allows parliament to broach issues that deal with how the executive carries out its responsibilities for the good of the nation, especially with regard to its financial responsibilities.

A Second Advancement: Chairing the Public Accounts Committee

Another outstanding issue that emerged from the benchmarking exercise was the protocol for appointing the chairperson of the Public Accounts Committee. In the past, the chairperson was always from the group of members forming the opposition. One of the other two remaining PAC members was also often from the opposition. However, when the former opposition became the government under Teburoro Tito, the member of parliament from South Tarawa, all PAC members (including the chairperson) subsequently came from the governing side (from 1994 to 2002).

In 2003, these trends shifted again when the opposition became the government under Anote Tong. One of the three PAC members was chosen from among the opposition members with the government's endorsement. At the benchmarking exercise, it was pointed out that most Commonwealth countries appoint the PAC chairperson from among the opposition members. However, members of the governing side noted that Kiribati, because of its very recent experience, was not yet ready to adopt this Commonwealth practice. No further argument ensued. There was a silent understanding and commitment by members (particularly senior members of the governing side) that adopting the widely accepted Commonwealth norm was the way forward. However, for the time being, the wish of the majority of members was taken into account.

During the review of the rules of parliament procedures, the rules governing the PAC chairperson affiliation were revisited. This time around, it was decided that the PAC chairperson should be elected from the opposition. Adopting this rule change has bolstered the reputation of Kiribati as being among the select set of small Pacific countries that are performing to the high standard of parliamentary democracy set by proven standards of Commonwealth parliaments.

Room for Improvement: Women in the House

As illuminated by the benchmarking exercise, a key area for further improvement in Kiribati's parliament is the low number of female parliamentarians. Among 44 elected parliamentarians, only 3 are women.

In the immediate term, we aim to address this issue by encouraging women to stand for parliamentary elections on the same and equal basis as given to men. We do not reserve a certain number of seats for women, nor are we prepared to pass legislation that would favor women candidates over their male counterparts. Such legislation would be, in our view, discriminatory in nature.

To date, women candidates have not advocated for special privileges over male contestants, as they feel that doing so would cast doubt on their abilities. Currently, women candidates' main disadvantage in getting elected is the fact that members of parliament have historically been male. However, Kiribati women are proving, and gaining recognition for, their capabilities in many positions in society that were traditionally held by men. Their performance has been as good as, if not better than, that of their male counterparts.

The low number of female parliamentarians cannot be rectified with a quick fix. For the time being, we will continue to encourage women to become members of parliament, but the burden is also on the electorate to make its own decisions. We believe that in taking this approach, we recognize the ability of women to become members of parliament but offer them the dignity inherent in being fairly elected.

Concluding Remarks: A Vision for Kiribati

The Parliament of Kiribati believes strongly in self-assessment so that it fulfills its role as a democratic institution that serves the needs and guarantees the rights of the people. We are continuously reminded of this role by our vision, which is to be an effective and transparent parliament that ensures respect for human rights, democracy, and good governance under a regime of the rule of law.

Equally important, we should serve our mission as elected servants of the people by performing our duty to be an open, transparent, and democratic parliament bound by the principles of good governance and accountability in order to effectively exercise the legislative, oversight, scrutinizing, and representative functions for the people of Kiribati. We are able to fulfill our mission because our parliament is supported by a strong constitution and by complementary rules of procedure as its backbone.

Annex 13A: CPA Recommended Benchmarks for Democratic Legislatures—Kiribati

Table 13A.1 lists the CPA benchmarks and notes whether they have been achieved.

Table 13A.1 Results of the Assessment Using the Commonwealth Parliamentary Association Benchmarks

CPA benchmark	Achieved Yes	No	Comments
I	General			
1	General			
1.1	Elections			
1.1.1 Members of the popularly elected or only house shall be elected by direct universal and equal suffrage in a free and secret ballot.	√		
1.1.2 Legislative elections shall meet international standards for genuine and transparent elections.	√		
1.1.3 Term lengths for members of the popular house shall reflect the need for accountability through regular and periodic legislative elections.	√		
1.2 Candidate Eligibility			
1.2.1 Restrictions on candidate eligibility shall not be based on religion, gender, ethnicity, race, or disability.	√		
1.2.2 Special measures to encourage the political participation of marginalized groups shall be narrowly drawn to accomplish precisely defined, and time-limited, objectives.		√	Three women are currently represented in parliament. If women are classified as a marginalized group, they are given equal opportunity to stand for elections, but there is currently no special treatment or measures to try and get them into parliament.
1.3 Incompatibility of Office			
1.3.1 No elected member shall be required to take a religious oath against his or her conscience in order to take his or her seat in the legislature.	√		
1.3.2 In a bicameral legislature, a legislator may not be a member of both houses.			Not applicable. Kiribati has a unicameral parliament.
1.3.3 A legislator may not simultaneously serve in the judicial branch or as a civil servant of the executive branch.			An exception is made for the attorney general.
1.4 Immunity			
1.4.1 Legislators shall have immunity for anything said in the course of the proceedings of legislature.	√		
1.4.2 Parliamentary immunity shall not extend beyond the term of office, but a former legislator shall continue to enjoy protection for his or her term of office.	√		
1.4.3 The executive branch shall have no right or power to lift the immunity of a legislator.	√		
1.4.4 Legislators must be able to carry out their legislative and constitutional functions in accordance with the constitution, free from interference.	√		

table continues next page

Table 13A.1 Results of the Assessment Using the Commonwealth Parliamentary Association Benchmarks *(continued)*

CPA benchmark	Achieved Yes	No	Comments
1.5 Remuneration and Benefits			
1.5.1 The legislature shall provide proper remuneration and reimbursement of parliamentary expenses to legislators for their service, and all forms of compensation shall be allocated on a nonpartisan basis.	√		
1.6 Resignation			
1.6.1 Legislators shall have the right to resign their seats.	√		
1.7 Infrastructure			
1.7.1 The legislature shall have adequate physical infrastructure to enable members and staff to fulfill their responsibilities.	√		
II Organization of the Legislature			
2 Procedure and Sessions			
2.1 Rules of Procedure			
2.1.1 Only the legislature may adopt and amend its rules of procedure.	√		
2.2 Presiding Officers			
2.2.1 The legislature shall select or elect presiding officers pursuant to criteria and procedures clearly defined in the rules of procedure.	√		
2.3 Convening Sessions			
2.3.1 The legislature shall meet regularly, at intervals sufficient to fulfill its responsibilities.	√		
2.3.2 The legislature shall have procedures for calling itself into regular session.	√		
2.3.3 The legislature shall have procedures for calling itself into extraordinary or special session.	√		Such procedures are provided for in the Constitution but have not been used.
2.3.4 Provisions for the executive branch to convene a special session of the legislature shall be clearly specified.	√		
2.4 Agenda			
2.4.1 Legislators shall have the right to vote to amend the proposed agenda for debate.	√		
2.4.2 Legislators in the lower or only house shall have the right to initiate legislation and to offer amendments to proposed legislation.	√		Private members bills are allowed, but money bills are not.
2.4.3 The legislature shall give legislators adequate advance notice of session meetings and the agenda for the meeting.	√		
2.5 Debate			
2.5.1 The legislature shall establish and follow clear procedures for structuring debate and determining the order of precedence of motions tabled by members.	√		
2.5.2 The legislature shall provide adequate opportunity for legislators to debate bills prior to a vote.	√		At times, the speaker restricts the debate because of time factors.

table continues next page

Table 13A.1 Results of the Assessment Using the Commonwealth Parliamentary Association Benchmarks *(continued)*

CPA benchmark	Achieved Yes	No	Comments
2.6 Voting			
2.6.1 Plenary votes in the legislature shall be public.	√		The exception is the appointment of the speaker and the president, which is done through secret ballot.
2.6.2 Members in a minority on a vote shall be able to demand a recorded vote.	√		
2.6.3 Only legislators may vote on issues before the legislature.	√		
2.7 Records			
2.7.1 The legislature shall maintain and publish readily accessible records of its proceedings.	√		
3 Committees			
3.1 Organization			
3.1.1 The legislature shall have the right to form permanent and temporary committees.	√		Usually, budget restrictions do not allow the formation of such committees.
3.1.2 The legislature's assignment of committee members on each committee shall include both majority and minority party members and reflect the political composition of the legislature.	√		
3.1.3 The legislature shall establish and follow a transparent method for selecting or electing the chairs of committees.	√		
3.1.4 Committee hearings shall be in public. Any exceptions shall be clearly defined and provided for in the rules of procedure.	√		
3.1.5 Votes of committee shall be in public. Any exceptions shall be clearly defined and provided for in the rules of procedure.	√		
3.2 Powers			
3.2.1 There shall be a presumption that the legislature will refer legislation to a committee, and any exceptions must be transparent, narrowly defined, and extraordinary in nature.	√		There are no committees as such. All legislation is considered by the committee of the whole house.
3.2.2 Committees shall scrutinize legislation referred to them and have the power to recommend amendments or amend the legislation.	√		There is no committee as such. All legislation is discussed by the committee of the whole house.
3.2.3 Committees shall have the right to consult and/or employ experts.	√		
3.2.4 Committees shall have the power to summon persons, papers, and records, and this power shall extend to witnesses and evidence from the executive branch, including officials.	√		
3.2.5 Only legislators appointed to the committee, or authorized substitutes, shall have the right to vote in committee.	√		Substitutes do not exist.

table continues next page

Table 13A.1 Results of the Assessment Using the Commonwealth Parliamentary Association Benchmarks (continued)

CPA benchmark	Achieved Yes	No	Comments
3.2.6 Legislation shall protect informants and witnesses presenting relevant information to commissions of inquiry about corruption or unlawful activity.		√	Commissions of inquiry are appointed and mandated by the president.
4 Political Parties, Party Groups, and Cross-Party Groups			
4.1 Political Parties			
4.1.1 The right of freedom of association shall exist for legislators, as for all people.	√		
4.1.2 Any restrictions on the legality of political parties shall be narrowly drawn in law and shall be consistent with the International Covenant on Civil and Political Rights.		√	The Constitution does not recognize political parties. Political parties are formed after the general election because forming them before an election is not required.
4.2 Party Groups			
4.2.1 Criteria for the formation of parliamentary party groups, and their rights and responsibilities in the legislature, shall be clearly stated in the rules.		√	
4.2.2 The legislature shall provide adequate resources and facilities for party groups pursuant to a clear and transparent formula that does not unduly advantage the majority party.		√	
4.3 Cross-Party Groups			
4.3.1 Legislators shall have the right to form interest caucuses around issues of common concern.	√		Members can form branches of the Commonwealth Parliamentary Association, and so forth.
5 Parliamentary Staff			
5.1 General			
5.1.1 The legislature shall have an adequate nonpartisan professional staff to support its operations, including the operations of its committees.		√	There are resource restrictions. Ideally, we need more staff members and more training for existing staff members.
5.1.2 The legislature, rather than the executive branch, shall control the parliamentary service and determine the terms of employment.		√	Staff appointment is carried out by the Public Service Commission (PSC).
5.1.3 The legislature shall draw and maintain a clear distinction between partisan and nonpartisan staff.		√	The staff of parliament comprises nonpartisan staff members who serve all members of parliament.
5.1.4 Members and staff of the legislature shall have access to sufficient research, library, and information, communication, and technology facilities.	√		At present, such facilities exist, but ideally this area needs improvement in terms of the number of research officers and computers. Currently, the whole complex has wireless Internet with three computers assigned for members of parliament to use.
5.2 Recruitment			
5.2.1 The legislature shall have adequate resources to recruit staff sufficient to fulfill its responsibilities. The rates of pay shall be broadly comparable to those in the public service.		√	Recruitment is carried out by the PSC.

table continues next page

Benchmarking and Self-Assessment for Parliaments • http://dx.doi.org/10.1596/978-1-4648-0327-7

Table 13A.1 Results of the Assessment Using the Commonwealth Parliamentary Association Benchmarks *(continued)*

CPA benchmark	Achieved		Comments
	Yes	*No*	
5.2.2 The legislature shall not discriminate in its recruitment of staff on the basis of race, ethnicity, religion, gender, disability, or, in the case of nonpartisan staff, party affiliation.		√	Selection and recruitment are carried out by the PSC.
5.3 Promotion			
5.3.1 Recruitment and promotion of nonpartisan staff shall be on the basis of merit and equal opportunity.	√		One would like to think that recruitment and promotion are based on merit, but recruitment and selection are the prerogative of the PSC.
5.4 Organization and Management			
5.4.1 The head of the parliamentary service shall have a form of protected status to prevent undue political pressure.		√	The clerk is appointed by the PSC, just like other civil servants.
5.4.2 Legislatures should, either by legislation or resolution, establish corporate bodies responsible for providing services and funding entitlements for parliamentary purposes and providing for governance of the parliamentary service.		√	Parliament is not autonomous and is subject to staff appointments made by the PSC, as well as to a budget that is provided by the Ministry of Finance.
5.4.3 All staff shall be subject to a code of conduct.		√	
III Functions of the Legislature			
6 Legislative Function			
6.1 General			
6.1.1 The approval of the legislature is required for the passage of all legislation, including budgets.	√		
6.1.2 Only the legislature shall be empowered to determine and approve the budget of the legislature.		√	The executive determines the budget to be provided to the legislature.
6.1.3 The legislature shall have the power to enact resolutions or other nonbinding expressions of its will.		√	The legislative budget is determined by the executive.
6.1.4 In bicameral systems, only a popularly elected house shall have the power to bring down government.	√		
6.1.5 A chamber where a majority of members are not directly or indirectly elected may not indefinitely deny or reject a money bill.	√		Kiribati does not have a second chamber. All members of parliament are elected.
6.2 Legislative Procedure			
6.2.1 In a bicameral legislature, there shall be clearly defined roles for each chamber in the passage of legislation.			Not applicable. Kiribati is a unicameral parliament.
6.2.2 The legislature shall have the right to override an executive veto.		√	If a law that is passed by parliament does not comply with the Constitution, the president can veto. Once this is done, parliament cannot override the veto.

table continues next page

Table 13A.1 Results of the Assessment Using the Commonwealth Parliamentary Association Benchmarks (continued)

CPA benchmark	Achieved		Comments
	Yes	No	
6.3 The Public and Legislation			
6.3.1 Opportunities shall be given for public input into the legislative process.	√		After the first reading of bills, there is a break before the second reading. At this time, members of parliament are allowed to leave and hold meetings in their constituency and seek the views of the people.
6.3.2 Information shall be provided to the public in a timely manner regarding matters under consideration by the legislature.	√		Members of parliament distribute the information themselves.
7 Oversight Function			
7.1 General			
7.1.1 The legislature shall have mechanisms to obtain information from the executive branch sufficient to exercise its oversight function in a meaningful way.	√		
7.1.2 The oversight authority of the legislature shall include meaningful oversight of the military security and intelligence services.			Not applicable. Kiribati does not have a military or intelligence service, but the legislature does have oversight of immigration and border control.
7.1.3 The oversight authority of the legislature shall include meaningful oversight of state-owned enterprises.	√		
7.2 Financial and Budget Oversight			
7.2.1 The legislature shall have a reasonable period of time in which to review the proposed national budget.	√		It has been proposed that a special parliamentary committee be established to look into the budget during its formulation stage, before its final presentation to parliament.
7.2.2 Oversight committees shall provide meaningful opportunities for minority or opposition parties to engage in effective oversight of government expenditures. Typically, the public accounts committee will be chaired by a member of the opposition party.		√	The chair of the Public Accounts Committee (PAC) is currently from the ruling party.
7.2.3 Oversight committees shall have access to records of executive branch accounts and related documentation sufficient to be able to meaningfully review the accuracy of executive branch reporting on its revenues and expenditures.	√		The PAC is able to do this.
7.2.4 There shall be an independent, nonpartisan supreme or national audit office whose reports are tabled in the legislature in a timely manner.	√		The Office of the Auditor General is appointed by the president.
7.2.5 The supreme or national audit office shall be provided with adequate resources and legal authority to conduct audits in a timely manner.	√		The chair of the PAC is allocated a vehicle for travel outside parliament during visits. One of the members of parliament reported that the auditor general actually had wanted more staff in his office.

table continues next page

Table 13A.1 Results of the Assessment Using the Commonwealth Parliamentary Association Benchmarks *(continued)*

CPA benchmark	Yes	No	Comments
	Achieved		
7.3 No Confidence and Impeachment			
7.3.1 The legislature shall have mechanisms to impeach or censure officials of the executive branch or express no confidence in the government.			The first part does not apply because the PSC deals with discipline for civil servants. The second part does apply (the legislature can express no confidence in the government).
7.3.2 If the legislature expresses no confidence in the government, the government is obliged to offer its resignation. If the head of state agrees that no other alternative government can be formed, a general election should be held.	√		This process can occur with an absolute majority.
8 Representational Function			
8.1 Constituent Relations			
8.1.1 The legislature shall provide all legislators with adequate and appropriate resources to enable the legislators to fulfill their constituency responsibilities.	√		The allowances of members of parliament have been increased throughout the years, but an entertainment allowance should be provided to cater for constituencies.
8.2 Parliamentary Networking and Diplomacy			
8.2.1 The legislature shall have the right to receive development assistance to strengthen the institution of parliament.	√		The United Nations Development Programme is currently finalizing a parliamentary strengthening project.
8.2.2 Members and staff of parliament shall have the right to receive technical and advisory assistance, as well as to network and exchange experience with individuals from other legislatures.	√		Parliament is currently on a twinning arrangement with the Australian Capital Territory Legislative Assembly.
IV Values of the Legislature			
9 Accessibility			
9.1 Citizens and the Press			
9.1.1 The legislature shall be accessible and open to citizens and the media, subject only to demonstrable public safety and work requirements.	√		
9.1.2 The legislature should ensure that the media are given appropriate access to the proceedings of the legislature without compromising the proper functioning of the legislature and its rules of procedure.	√		Proceedings are broadcast on live radio.
9.1.3 The legislature shall have a nonpartisan media relations facility.		√	
9.1.4 The legislature shall promote the public's understanding of the work of the legislature.		√	The absence of a media officer prevents such promotion; however, the live radio broadcast is a way of informing the public about the activities of parliament.
9.2 Languages			
9.2.1 Where the constitution or parliamentary rules provide for the use of multiple working languages, the legislature shall make every reasonable effort to provide for simultaneous interpretation of debates and translation of records.			Not applicable. Debate is done in the Kiribati language only.

table continues next page

Table 13A.1 Results of the Assessment Using the Commonwealth Parliamentary Association Benchmarks *(continued)*

| | | Achieved | | |
CPA benchmark		Yes	No	Comments
10	Ethical Governance			
10.1	Transparency and Integrity			
10.1.1	Legislators should maintain high standards of accountability, transparency, and responsibility in the conduct of all public and parliamentary matters.	√		
10.1.2	The legislature shall approve and enforce a code of conduct, including rules on conflicts of interest and the acceptance of gifts.		√	A leadership code of conduct was defeated the last time it was introduced.
10.1.3	Legislatures shall require legislators to fully and publicly disclose their financial assets and business interests.		√	There is a register of members' interests. However, the wealth of members of parliament is not exorbitant, so there would not be much to declare.
10.1.4	There shall be mechanisms to prevent, detect, and bring to justice legislators and staff engaged in corrupt practices.	√		There are normal provisions under the law.

Annex 13B: Application of the Inter-Parliamentary Union Toolkit in Kiribati

A self-assessment was also conducted using the Inter-Parliamentary Union's Self-Assessment Toolkit for Parliaments. The toolkit includes a rating scale, from one to five, where five is the highest score possible. The results of the self-assessment are shown table 13B.1.

Table 13B.1 Results of the Assessment Using the Inter-Parliamentary Union Toolkit

Question		Rating	Comments
The representativeness of parliament			
1.1	How adequately does the composition of parliament represent the diversity of political opinion in the country (for example, as reflected in votes for the respective political parties)?	4	This question applies to a system that does not exist in Kiribati. Kiribati has a homogeneous society, where the party system is not encouraged and where parties are formed only after the general elections.
1.2	How representative of women is the composition of parliament?	1	There are more women in the executive branch, acting as chief executive officers and in other roles.
1.3	How representative of marginalized groups and regions is the composition of parliament?	5	There are no marginalized groups. The Banabans and Ocean Islanders also have their representatives in parliament.
1.4	How easy is it for a person of average means to be elected to parliament?	3	No deposit is required for a Banaban candidate. Wealthy candidates are able to give money to their electorate, whereas poor candidates cannot do so.

table continues next page

Table 13B.1 Results of the Assessment Using the Inter-Parliamentary Union Toolkit *(continued)*

Question	Rating	Comments
1.5 How adequate are internal party arrangements for improving imbalances in parliamentary representation?	n.a.	No party system exists until after the elections.
1.6 How adequate are arrangements for ensuring that opposition and minority parties or groups and their members can effectively contribute to the work of parliament?	4	There is equal chance for all groups to contribute. However, arrangements are not perfect and cannot please everyone at all times.
1.7 How conducive is the infrastructure of parliament and its unwritten mores to the participation of women and men?	4	Equal opportunities exist for both men and women.
1.8 How secure is the right of all members to freely express their opinions, and how well are members protected from executive or legal interference?	5	
1.9 How effective is parliament as a forum for debate on questions of public concern?	5	
Parliamentary oversight of the executive		
2.1 How rigorous and systematic are the procedures whereby members can question the executive and secure adequate information from it?	5	
2.2 How effective are specialist committees in carrying out their oversight function?	5	The Public Accounts Committee is very effective.
2.3 How well is parliament able to influence and scrutinize the national budget through all its stages?	4	Parliament is not involved in the formulation stage.
2.4 How effectively can parliament scrutinize appointments to executive posts and hold their occupants to account?	4	
2.5 How far is parliament able to hold nonelected public bodies to account?	4	This question applies to statutory bodies.
2.6 How far is parliament autonomous in practice from the executive (for example, through control over its own budget, agenda, timetable, personnel, and so forth)?	1 5	Parliament does not have control over its budget and personnel. Parliament has full control over its agenda and timetable.
2.7 How adequate are the numbers and expertise of the professional staff to support members, individually and collectively, in the effective performance of their duties?	2 4	In the overall staffing structure, ideally more staff members are needed. However, the expertise of current staff members is rated highly.
2.8 How adequate are the research, information, and other facilities available to members and their groups?	3	Ideally, more staff members are needed
Parliament's legislative capacity		
3.1 How satisfactory are the procedures for subjecting draft legislation to full and open debate in parliament?	4	
3.2 How effective are committee procedures for scrutinizing and amending draft legislation?	n.a.	All legislation is discussed in the committee of the whole house.

table continues next page

Table 13B.1 Results of the Assessment Using the Inter-Parliamentary Union Toolkit *(continued)*

Question	Rating	Comments
3.3 How systematic and transparent are the procedures for consultation with relevant groups and interests in the course of legislation?	n.a.	All debate and discussion is carried out in the committee of the whole house. However, members of parliament are allowed to return to and seek the views of their constituencies after the first reading of the bill.
3.4 How adequate are the opportunities for individual members to introduce draft legislation?	5	There are no restrictions except on money bills and related matters.
3.5 How effective is parliament in ensuring that legislation enacted is clear, concise, and intelligible?	5	During the committee stage, editorial amendments can be proposed to amend any unclear provisions.
3.6 How careful is parliament in ensuring that legislation enacted is consistent with the constitution and the human rights of the population?	5	The president can veto legislation if it does not comply with the Constitution. Human rights issues go through a referendum and the attorney general will include an explanatory note that parliament has complied with human rights standards.
3.7 How careful is parliament in ensuring a gender-equality perspective in its work?	3	Words such as *man* are used to represent both sexes.
Transparency and accessibility of the legislature		
4.1 How open and accessible to the media and the public are the proceedings of parliament and its committees?	4	Media are allowed into parliament, whereas their admittance to committee meetings is left to the discretion of the chair.
4.2 How free from restrictions are journalists in reporting on parliament and the activities of its members?	4	Journalists have freedom in reporting on parliament but are liable for slanderous comments.
4.3 How effective is parliament in informing the public about its work through a variety of channels?	2	Parliament does not have a media officer who can act in this regard.
4.4 How extensive and successful are attempts to interest young people in the work of parliament?	3	Attempts to instill interest have included school visits to parliament, Commonwealth Day celebrations, and lessons on parliamentary democracy as part of the educational curriculum.
4.5 How adequate are the opportunities for electors to express their views and concerns directly to their representatives, regardless of party affiliation?	4	Members work for all electors, and it would be extremely detrimental to a member's interests not to listen to everyone's views.
4.6 How user-friendly is the procedure for individuals and groups to make submissions to a parliamentary committee or commission of inquiry?	4	Only the Public Accounts Committee is able to hold public hearings, and press releases are issued to invite public submissions. Commissions of inquiry are mandated by the president.
4.7 How much opportunity do citizens have for direct involvement in legislation (for example, through citizens' initiatives and referenda)?	4	Members of parliament have the opportunity to go back to their constituencies to seek the views of their electorates after the first reading stage of bills. These views are then brought back to the house during the second reading debate stage. A referendum is carried out whenever there is a human rights issue as contained in the Constitution.

table continues next page

Benchmarking and Self-Assessment for Parliaments • http://dx.doi.org/10.1596/978-1-4648-0327-7

Table 13B.1 Results of the Assessment Using the Inter-Parliamentary Union Toolkit *(continued)*

Question	Rating	Comments
The accountability of parliament		
5.1 How systematic are arrangements for members to report to their constituents about their performance in office?	4	Members of parliament usually have two weeks before a session and immediately after the first reading stage of legislation to report to their constituents.
5.2 How effective is the electoral system in ensuring the accountability of parliament, individually and collectively, to the electorate?	4	Election is carried out on the notion that the system will be free and fair. In recent years, voters have expected more accountability from members of parliament.
5.3 How effective is the system for ensuring the observance of agreed codes of conduct by members?	1	There is no provision for having a code of conduct in the constitution.
5.4 How transparent and robust are the procedures for preventing conflicts of financial and other interests in the conduct of parliamentary business?	1	Such procedures are not provided for in the standing orders. Kiribati is a small country, and everyone knows each other (personally and in terms of each other's wealth).
5.5 How adequate is the oversight of party and candidate funding to ensure that members preserve independence in the performance of their duties?	n.a.	Parties are formed only after the election.
5.6 How publicly acceptable is the system whereby members' salaries are determined?	4	A tribunal is appointed by the chair of the Public Service Commission, and the tribunal's report is tabled in the Cabinet so that a bill is tabled in parliament for debate.
5.7 How systematic are the monitoring and review of levels of public confidence in parliament?	1	This matter is seen as the role of the executive. If parliament is not invaded, then public confidence must be good.
Parliament's involvement in international policy		
6.1 How effectively can parliament scrutinize and contribute to the government's foreign policy?	1	Scrutiny of and contribution to the government's foreign policy is purely the role of the executive.
6.2 How adequate and timely is the information available to parliament about the government's negotiating positions in regional and international bodies?	1	Information is released only through parliamentary questioning. The government is not obliged to report on its negotiating positions.
6.3 How much can parliament influence the binding legal or financial commitments made by the government in international forums, such as the United Nations?	1	
6.4 How effective is parliament in ensuring that international commitments are implemented at the national level?	1	
6.5 How effectively can parliament scrutinize and contribute to national reports to international monitoring mechanisms and ensure follow-up on their recommendations?	1	
6.6 How effective is parliamentary monitoring of the government's development policy, whether as "donor" or "recipient" of international development aid?	1	

table continues next page

Table 13B.1 Results of the Assessment Using the Inter-Parliamentary Union Toolkit *(continued)*

Question	Rating	Comments
6.7 How rigorous is parliamentary oversight of the deployment of the country's armed forces abroad?	n.a.	Kiribati does not have armed forces.
6.8 How active is parliament in fostering political dialogue for conflict resolution, both at home and abroad?	n.a.	The government uses the churches for conflict resolution purposes.
6.9 How effective is parliament in interparliamentary cooperation at the regional and global levels?	5	The Parliament of Kiribati is linked to the Commonwealth Parliamentary Association. It participates in the Inter-Parliamentary Union and United Nations Development Programme Institutional Strengthening Project.
6.10 How much can parliament scrutinize the policies and performance of international organizations such as the United Nations, the World Bank, and the International Monetary Fund, to which its government contributes financial, human, and material resources?	1	
6.11 Is there any special committee or entity in parliament with a specific mandate to monitor and follow up on matters relating to the United Nations, and if so, which body and what mandate does the body have?	1	
6.12 Are members of parliament included, as a matter of course, in government delegations to the United Nations General Assembly or to other multilateral forums?	1	This procedure would be good to implement.
6.13 Do ministers report to parliament on progress in international negotiations?	1	They do so only if they are questioned. Ministers are not required to report back to parliament.
6.14 Are national reports to international monitoring mechanisms of international conventions and agreements reviewed, debated, and approved in parliament before submission and are recommendations from such bodies tabled in parliament?	1	

Note: 5 = very high or very good; 4 = high or good; 3 = medium; 2 = low or poor; 1 = very low or very poor; n.a. = not applicable.

Note

1. Also attached, as annex 13B, is the parliament's self-assessment using the Inter-Parliamentary Union toolkit.

CHAPTER 14

Assessing Parliament Using the CPA Benchmarks: A Personal Perspective from Bermuda

Jennifer Smith

Introduction

I will first explain the background against which Bermuda's parliament developed and, in doing so, draw attention to its particular strengths and weaknesses. I do this from the vantage point of having served as a backbencher, shadow minister, opposition leader, premier, deputy speaker, and most recently as minister of education.

Bermuda has the oldest parliament in the Western Hemisphere. It is a bicameral legislature, exercising parallel functions to those of the House of Commons and the House of Lords in the United Kingdom. Bermuda's parliament first met in 1620 and is currently housed in a building that began as a simple two-story edifice in 1819. Parliament moved into the building in 1826 and added a clock tower and Florentine façade in 1893. Many additions have been made to the building over the years, including the most recent to provide elevator access to the second story. It is an attractive building, but wholly inadequate to the current needs of parliament in terms of infrastructure.

The House of Assembly consists of 36 elected members—19 from the One Bermuda Alliance and 17 from the Progressive Labour Party. Situated in a different building is the upper house, or Senate, with 11 members. Five are from the government, three are from the opposition, and three are independent.

The presiding officers of both houses—the speaker in the House of Assembly and the president in the Senate—are elected by their peers in their respective chambers. Their roles are similar: to preside over meetings, regulate debate, arbitrate on procedural matters, make decisions on points of order, and give rulings when and where necessary.

The author would like to acknowledge the assistance of Shernette Wolffe, clerk of the Bermuda House of Assembly, in writing this chapter.

The speaker is usually elected from the majority party, whereas the president of the Senate has always been elected from among the independent senators. Once elected, the speaker renounces party affiliation and does not participate in any of the debates. A deputy speaker and a vice president are also elected by their peers to assist the presiding officers in the event of illness or absence or to provide relief during lengthy debates.

One difference between the two houses is that the president of the Senate can contribute during deliberations and vote alongside his or her peers. The speaker, however, cannot take part in House of Assembly debates and can vote only in the event of a tie. All proceedings of both houses are open to the public and are also broadcast gavel to gavel on the radio.

The House of Assembly meets once a week on Fridays, and the Senate meets once a week on Wednesdays. During the budget debate, meetings are held three times a week (Mondays, Wednesdays, and Fridays) over a period of three weeks to ensure that all discussion is completed and the relevant legislation is passed and assented to by the governor before the March 31 financial year deadline.

Both houses of the legislature meet together only on occasions of great significance such as the convening of parliament, the 25th anniversary of Bermuda's constitution (1993), the 375th anniversary of Bermuda's parliament (1995), or events paying tribute to significant members who have died while in office.

Results of the Benchmark Exercise

In Bermuda, expenditure on parliament has always been viewed as expenditure on the members of parliament (MPs), who—it is widely viewed—should serve for free. When the Commonwealth Parliamentary Association (CPA) first began looking at minimum benchmarks for democratic legislatures, Bermuda's MPs were immediately interested because they needed an independent platform on which to base the changes that were needed and to educate civil servants and the public about the basic needs and services of parliament.

Bermuda's MPs began by reviewing the parliament's rules. The prime instrument used in this process was the CPA's Eastern Caribbean template, a modernized set of common standing orders produced in 2007 with technical expertise provided by the Ontario Legislative Assembly for use by the nine small parliaments and legislatures of the Organisation of Eastern Caribbean States. The Rules and Privileges Committee set up a subcommittee consisting of Opposition Member John Barritt and myself. Thus began a process that had not been undertaken for more than 20 years. As of this writing, the new standing orders have been provisionally accepted by the House of Assembly and are now in use. One outstanding matter still has to be resolved by the Standing Orders Committee before the orders are finally ratified.

Bermuda hosted a CPA benchmarks seminar to go through the process of measuring the parliament against the benchmarks. This exercise allowed parliamentarians to see for themselves that Bermuda was not up to scratch. I can assure you that this finding surprised them because we like to think of ourselves as first in all areas. Also, with one of the oldest parliaments, they thought Bermuda would be the most democratic. Not so.

The exercise showed that Bermuda fell short when it came to benchmarks 1.7.1, 5.1.2, 5.2.1, 6.1.2, and 8.1.1. Specifically, the following fundamental areas needed improvement:

- Adequate and proper facilities
- Sufficient qualified staff
- Operating budget adequate to the need of serving both parliamentarians and the public

Benchmark 1.7.1

Benchmark 1.7.1 states, "The legislature shall have adequate physical infrastructure to enable members and staff to fulfill their responsibilities."

Legislating is parliament's primary responsibility. As an isolated island (the northernmost archipelago in the Atlantic Ocean), this responsibility could be made so much easier in Bermuda with the use of technology. However, there is no room to provide a parliamentary library and research service for MPs without undergoing extensive building renovation. In the interim, MPs need not only to access the web but also to access information to assist them in the chamber for debates, in committees, and in caucus. Other parliaments post their debates, laws, and research materials online the next day, and as a result MPs can access this information from their seats.

MPs should also be able to use e-mail and other web-based technology to communicate with their constituents and colleagues around the world. Their ability to do so cannot be based on personal, business, or political affiliation.

Of course, parliamentary staff members have access to the Internet—after all, Bermuda is a leader in e-commerce and telecommunications—and they help MPs as much as they can. But the fundamental need for MPs to have Internet access required us first to wire the chamber so that members would be able to use laptops and then to amend our rules, even though they had recently been rewritten on the basis of the CPA's Eastern Caribbean template, so that use of technological communications tools (including BlackBerry devices) would no longer violate the rules.

Benchmark 6.1.2

Benchmark 6.1.2 states, "Only the legislature shall be empowered to determine and approve the budget of the legislature."

Until recently, the parliamentary budget was under the control of a civil service permanent secretary, then the minister of justice, and then the Cabinet. Thus, having budgeted for the introduction of a Hansard, a multiyear project that was in its final phase, MPs had the indignity of having the permanent secretary arbitrarily delete funding because "it hadn't been spent." Of course, Cabinet members reinstated the funding because the Hansard was something they wanted and had made one of their own "Throne Speech" promises. Now parliament is a nonministry department, and an organizational review of parliament is being considered to further modernize its operations so that we meet the benchmarks.

Benchmarks 5.1.2 and 5.2.1

Benchmark 5.1.2 states, "The legislature, rather than the executive branch, shall control the parliamentary service and determine the terms of employment."

And benchmark 5.2.1 states, "The legislature shall have adequate resources to recruit staff sufficient to fulfill its responsibilities."

When parliament was still under ministerial control, a nonfunctioning staff member was transferred. The situation was this: The clerk had established a need for an additional staff member, but she had to put this request before the Ministry of Justice permanent secretary, who, instead of looking at the qualifications and experience needed for the post, transferred a ministry staff member to the post. It was clear almost immediately that this person was not a good fit, but it took the clerk more than a year to rectify the situation. The clerk had to start over and find someone who would be a good fit for the post.

Benchmark 8.1.1

Benchmark 8.1.1 states, "The legislature shall provide all legislators with adequate and appropriate resources to enable the legislators to fulfill their constituency responsibilities."

Since the 2007 general election, the makeup of Bermuda's parliament has changed dramatically. It began with 22 government members and 14 opposition members. Since then, first one and then two opposition members left to sit as independents. Then three opposition members left to sit as a new party called the Bermuda Democratic Alliance. Since May 2011, the United Bermuda Party has been in a state of flux, first voting to merge with the Bermuda Democratic Alliance to form the One Bermuda Alliance and then splitting so the remaining members of the United Bermuda Party were outside of the One Bermuda Alliance, which took over as the official opposition. At the time when this chapter was written, changes were still occurring.

The opposition party changes affect the benchmark on the provision of resources to members. Historically, Bermuda's parliament provided no more than minimum services for members—no postage stamps, no transportation, and no secretarial services, although there is free local telephone service. The ability of parliamentarians to carry out their legislative and constitutional functions depends almost entirely on the political parties. Now, with the current party makeup, we can no longer rely on political parties to provide for their members and have to ensure that all members are able to communicate with their constituents and carry out their representative duties.

The government has expressed its intention to raise members' salaries, but in the current economic climate, it has not been able to move forward with this intention. It is unfortunate, but true, that the Bermudian public perceives any funding given to parliament to be politicians giving to themselves—a perception that can be eradicated through education and training.

Although I am proud of the parliament and its longevity, there is a clear need identified by all parties concerned for both improvement and strengthening.

New Benchmarks

Using the 2006 CPA Recommended Benchmarks as a template (CPA 2006), a group of clerks representing the CPA region of the Caribbean, Americas, and Atlantic met in Barbados on March 8 and 9, 2011, and drafted regional versions of the CPA benchmarks.

The group felt that worldwide parliamentary standards are continuously evolving and that all parliaments can be sources of valuable innovations regardless of their size or age. Additionally, the group felt it important to develop benchmarks on the basis of the unique traditions and parliamentary practices of the Caribbean region.

A number of significant additions, omissions, and modifications were therefore proposed to the 2006 Recommended Benchmarks. Finally the draft Caribbean benchmarks were finalized and adopted at the 36th CPA Regional Conference of the Caribbean, Americas, and Atlantic held in Grenada in June 2011. In the document (CPA 2011), significant new benchmarks were either added or modified.

Benchmark 1.7.2

Benchmark 1.7.2 states, "Members shall be entitled to have adequate office accommodation, with modern amenities throughout their term in office."

Bermuda's MPs do not have office space to meet their constituents or representatives of various organizations relevant in their constituencies. Currently, members use the existing committee rooms to meet their constituents.

Benchmark 1.7.4

Benchmark 1.7.4 states, "Elected members shall be provided with state-funded offices in their constituencies."

Some sister island countries to the south, such as Barbados, Jamaica, and Trinidad and Tobago, provide members not only with state-funded offices but also with constituency assistants. Further, constituency development funds are used to finance constituency projects. Currently, members are not provided with funding to assist their constituents, and they are expected to fund programs and initiatives out of their own pockets. Parliamentary staff members even have to send out letters on behalf of MPs who want to congratulate their constituents.

Benchmark 5.1.1

With respect to parliamentary staff, a revised benchmark 5.1.1 states, "The legislature shall have adequate professional staff to support its operations, including the operations of its committees, but where applicable, members are entitled to choose their own personal staff."

Bermuda's parliament has always been served by a very small complement of staff members. The majority of the parliamentarians were wealthy businessmen, and their private staffs carried out most functions on their behalf. Parliamentary service was only minimal. The idea of having a qualified and adequate professional staff with certain areas of expertise was not even touted. Currently, parliament is staffed by seven employees, and as a result, they have to don many hats.

Clearly Bermuda cannot continue on this path, and some members have now embraced the idea for reform after being apprised of the CPA benchmarks. I am hopeful that a review and assessment of our legislative practices will happen in the near future.

Benchmark 6.1.6

Finally, benchmark 6.1.6 states, "Ministries and departments shall transmit bills and other documents for parliamentary action to the clerk of each house in electronic form for timely distribution to members. Hard copies of such documents, if required, shall also be transmitted in accordance with the established practice."

Bermuda's parliament is making some strides in encouraging a paperless environment. As a matter of fact, one of the newly revised standing orders uses the same wording as benchmark 6.1.6.

Conclusion

To conclude, although Bermuda's parliament has some distance to go to meet certain benchmarks, it has made gradual steps toward reaching the ultimate goal. To date, the parliament has a Hansard, of which we are proud. Also, the parliamentary website was fully launched in July 2013, and now parliamentarians have individual parliament.bm e-mail addresses to which their constituents can send their concerns. A comprehensive e-mail and Internet policy has been written, which members must sign before having full access to their parliament.bm email address.

References

CPA (Commonwealth Parliamentary Association). 2006. "Recommended Benchmarks for Democratic Legislatures." CPA, London. http://wbi.worldbank.org/wbi/Data/wbi/wbicms/files/drupal-acquia/wbi/Recommended%20Benchmarks%20for%20Democratic%20Legislatures.pdf.

———. 2011. "Recommended Benchmarks for the CPA Caribbean, Americas, and Atlantic Region Democratic Legislatures." CPA, London. http://www.cpa-caaregion.org/media/get_media.php?mediaid=caa4fafb-a31.

CHAPTER 15

The African Parliamentary Index: Case Studies

Rasheed Draman

Introduction

As noted in chapter 9, the African Parliamentary Strengthening Program (APSP) for Budget Oversight is a five-year capacity strengthening program for seven partner parliaments: Benin, Ghana, Kenya, Senegal, Tanzania, Uganda, and Zambia. The program, funded by the Canadian International Development Agency and implemented by the African Program of the Parliamentary Centre, supports the seven partner parliaments in developing and implementing strategies to strengthen their overall role and engagement in the national budget process.

The program is premised on the fact that the budget process is a key area of focus for parliaments and relates closely to poverty reduction. Because government budgets are concerned about resource allocation that affects the lives of citizens, equipping elected representatives with the requisite tools to facilitate their role in the budget process is critical.

The Parliamentary Centre would like to thank the leadership of all seven partner parliaments of the African Parliamentary Strengthening Program for enabling evidence-gathering visits by Parliamentary Centre staff members and consultants. We are most grateful to the members of parliament and staff members who took time from their busy schedules to fully participate in the self-assessment process. The contribution of the independent country assessors and members of the support staff who ensured that participants understood the African Parliamentary Index concept and assigned scores is highly appreciated. We are especially grateful to the civil society organizations that responded to our invitations with alacrity and actively participated in the assessment. They provided a different perspective, which validates the parliamentarians' self-assessment. To the dedicated staff members of the Parliamentary Centre, who worked tirelessly to ensure the completion and launch of the report, we say well done. We are particularly grateful for the contribution of Elvis Otoo, a former monitoring and evaluation governance expert at the Parliamentary Centre, who developed the concept for the index and ensured the buy-in of partner parliaments. We also commend the contribution of Cynthia A. Arthur and Issifu Lampo for reviewing all the country reports submitted by the independent country assessors and for providing considerable editorial input. Lastly, we thank Gifty Adika, the coordinator of operations for Africa programs, and her staff for managing and seamlessly executing the administrative process.

The African Parliamentary Index (API) was designed by the Parliamentary Centre to provide a standard and simplified system for assessing the performance of parliaments in Africa, especially the seven APSP partner parliaments. The assessment process was both broad based and participatory across all APSP partner countries. Key stakeholders, including members of parliament (MPs), the parliamentary staff, independent research institutions, civil society organizations (CSOs), and university academics, participated in the assessment process, thus helping to ensure legitimacy and country ownership of the final outcome.

The findings, summarized here, point to participating parliaments' obvious capacity strengths as well as demonstrable weaknesses. The hope is that these parliaments find this information useful in designing their strategic programming and capacity-strengthening plans.

On the basis of what is believed to be good parliamentary practice in improving democracy and effective governance in Africa, the seven APSP countries were assessed under five core areas: representation, legislation, parliamentary oversight, institutional capacity, and transparency and integrity. See chapter 9 for a detailed description of the scope, approach, and methodology used.

Representation

Parliaments embody the will of the citizens and thus provide the space to express that will. They provide a forum where issues of local and national importance are raised and debated and where these debates are then translated into policies. Effective representation requires MPs to continually interact with their constituents to understand their views and perspectives and to use various legislative or parliamentary processes, such as questions, motions, resolutions, and other oversight mechanisms, to bring these views to the attention of implementing institutions to consider and redress. Overall, the effectiveness of the MPs' representational role and, indeed, parliament as a whole, depends to a large extent on the quality of the interaction between constituents and MPs. The API looks at how accessible the legislature is to the public and what efforts the legislature makes to help the public to understand its role.

All seven parliaments indicated the relative importance of their representation function by assigning high weights to accessibility. The challenge was how to harness existing capacity to make the parliaments more accessible to the public. In terms of parliamentary capacity to represent its people, figure 15.1 shows how the various parliaments fared.

The Tanzanian parliament's self-assessment indicated the highest capacity to represent its people. It was followed by the parliaments of Uganda, Ghana, Senegal, Zambia, and Kenya. Benin's parliament had the least capacity to meet the expectation of its people in terms of representation. In assigning a weight of 8.3 out of a maximum rate of 10 to the Tanzanian parliament's ability to represent it people, parliamentary assessors indicated that the legislature is accessible to citizens and the media, has a nonpartisan media center, and has mechanisms to promote citizens' understanding of its work. Tanzania's parliamentary

Figure 15.1 Weighted Capacity Ratio: Representation

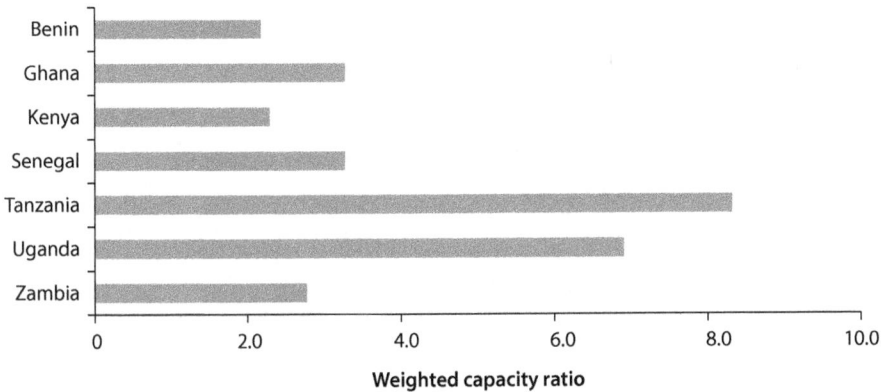

Source: African Parliamentary Index, Parliamentary Centre.

assessors also reported one negative thing: information flow to the public is often not as timely as they would have wanted. (Tanzanian CSOs disagreed with parliament's self-assessment and assigned a weighted capacity score of only 5.6.)

In Benin's case, parliamentary assessors thought the legislature was not as open to citizens and the media as perhaps it could be and that mechanisms for promoting citizens' understanding of the legislature were nonexistent, even though some attempts have been made to create public awareness of parliament's work. Benin participants assigned a weighted capacity score of only 2.2.

Legislation

Lawmaking is a core function of the legislature, typically vested by the constitution of a country but sometimes vested by an act of parliament. Whether a bill is initiated by the executive or a private member, it is the legislature's responsibility to consider such a bill and pass it into law when a majority of MPs support it. The legislature's control of the purse is expressed in its power to pass the appropriations act, which allocates financial resources to the executive and other institutions of the state. The API assesses the factors that affect the legislature's effectiveness in executing its legislative function. The assessment also covers parliaments' legal mandate, with emphasis on the source of the authority of the legislative power, whether the public has input into the legislative process, and whether the legislature has a mechanism to monitor the effect of laws passed.

Figure 15.2 presents how the seven countries fared in the self-assessment. Kenya's parliament scored the highest in this indicator with a weighted capacity average of 9.3, followed by Benin and Senegal with 8.1 each and then Ghana, Zambia, and Uganda. Tanzania had the least capacity in this area of assessment (a score of 6.2). The Kenya parliamentary assessors indicated the relative importance of their lawmaking function and assigned a high weighting coefficient of

Figure 15.2 Weighted Capacity Ratio: Legislation

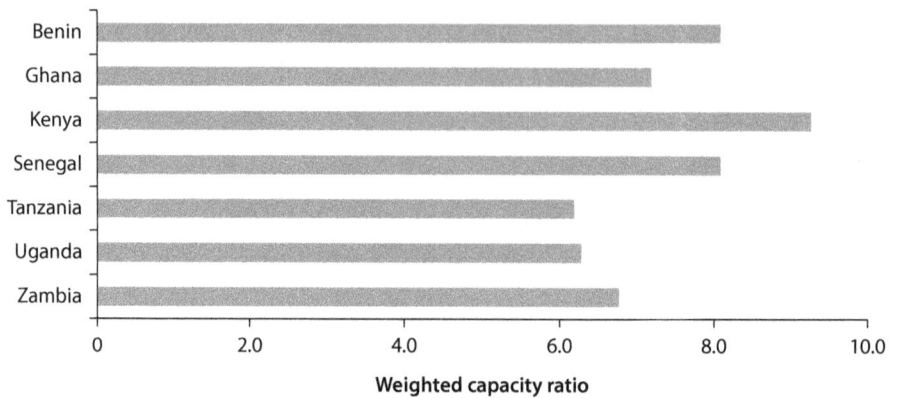

Source: African Parliamentary Index, Parliamentary Centre.

13 compared with that assigned to the other indicators. Because of the high importance attached to the legal mandate of parliament, the power of the legislature to make laws, including the appropriations act, is enshrined in Kenya's constitution, and adequate provisions exist for citizens to provide input into the legislation process, even though their input is not backed by legislation. Participants also indicated that the legislature can amend the appropriations bill only with the consent of the minister of finance and executive. Some mechanisms exist for tracking enacted legislations, but this area needs adequate resourcing. Kenyan CSOs, in their validation assessment of this indicator, perceived parliament to have more power and capacity than parliament thought its legal mandate covered. CSOs assigned an average weight of 3.4, which resulted in a computed capacity weight of 11.

In the case of Tanzania, which had the least capacity in this indicator, the legislature's inability to amend the appropriations bill was a source of concern.

Power of the Purse

The financial function is one of the legislature's major responsibilities. Also referred to as the "power of the purse" in parliamentary parlance, it implies the legislature controls the resources and finances of the state and, therefore, its responsibility to disburse such resources. In most countries, the legislature must approve taxes and also determine how those taxes are expended. Thus, the financial function transcends the mere allocation of funds to encompass a general understanding of economic indicators and ways decisions of the legislature, such as tax increases and the imposition of levies, affect economic activity generally. This indicator assesses the strengths of a legislature in executing this financial function. In particular, it examines the legislature's involvement in the budget process and whether the legislature can change proposals submitted by the executive. It further assesses availability of technical expertise to the legislature.

Figure 15.3 Average Weighted Capacity Ratio: Financial Function

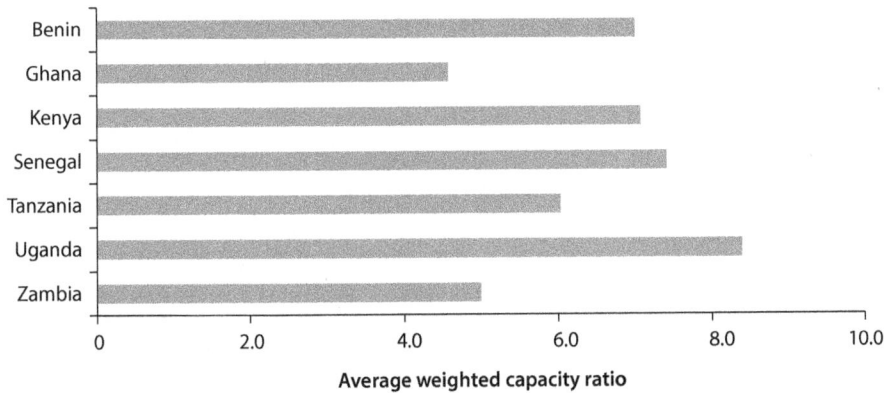

Source: African Parliamentary Index, Parliamentary Centre.

Of particular interest is the existence of a budget act and a budget office, as well as the ex-post parliamentary review mechanisms. Figure 15.3 presents average weighted capacity of all three subcategories of the financial function of parliament: (a) budget review and hearing, (b) budget act and budget office, and (c) periodic review of the budget.

Uganda had the highest capacity in this category (a score of 8.4), followed by Senegal, Kenya, Benin, Tanzania, and Zambia. Ghana, with a 4.6, scored the least in terms of its capacity to perform financial functions. Because this indicator assesses the existence of the budget act and office, it is not surprising that Uganda and Kenya are among the top performers. Uganda, for instance, has a budget act that clearly defines the role for the legislature in the budget process. This fact is confirmed by CSOs, which nevertheless call for a review of the act to enhance oversight of supplementary expenditure ceilings and match and harmonize the planning framework with other legal frameworks. Kenya, in contrast, has the Fiscal Management Act, which provides for a more assertive role by parliament in overseeing the national budget and which also established a budget office.

In the case of the two francophone countries, Benin and Senegal, parliamentary assessors were of the view that their "organic laws" for public finance adequately regulated the legislature's role in the budget process. Though the laws do not recommend setting up a budget office, there is a budget and finance committee within the legislatures of both countries, which, in part at least, undertakes the work of a budget office. It is on this basis that the high capacity scores of 7.4 and 7.0 were awarded to Senegal and Benin, respectively.

Parliamentary Oversight

Effective parliamentary oversight is one of the tools used by the legislature to maintain a balance of power among the three arms of government and to assert the interests of ordinary citizens against the decisions of the executive.

Figure 15.4 Average Weighted Capacity Ratio: Oversight Function

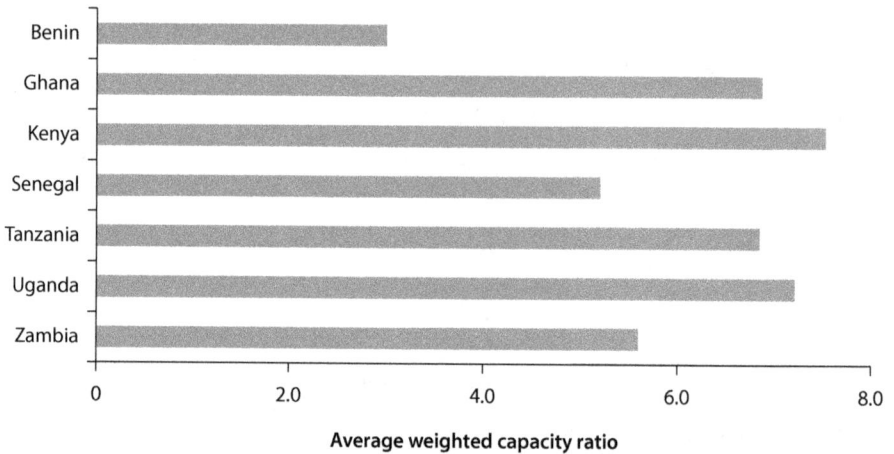

Source: African Parliamentary Index, Parliamentary Centre.

The committee system is a key tool for oversight because it allows the break-down of parliamentary oversight into small chunks that are based on themes and subject matter. Hence, the committee system allows parliamentarians to develop expertise and to conduct thorough examinations of proposed legislation. With a well-functioning committee system, executive policies, actions, and expenditure are subject to greater scrutiny and oversight. This indicator assesses the existence and effectiveness of relevant oversight parliamentary committees, their powers, and the resources available to them. It also examines the existence and effectiveness of a public accounts committee (PAC) and an auditor general, looks at the powers and responsibilities of the two bodies, and considers whether they have requisite resources to enable them to deliver. Figure 15.4 illustrates each of the seven countries' capabilities.

These scores represent the average score for three subindicators: the existence and functions of the oversight committees in general, PACs in particular, and a parliamentary auditor. The assessment by partner-country parliaments revealed Kenya has the most capacity to execute its oversight functions, followed by Uganda, Ghana, Tanzania, and Zambia.

Again confirming the institutional governance differences, the two franco-phone countries, Senegal and Benin—which have a different oversight setup from the one anticipated by the index—suffered lower capacity ratings of 5.2 and 3.0, respectively. The reason for their lower scores is that they have no dedicated parliamentary committee responsible for the oversight of public accounts, as exists in anglophone countries. Furthermore, rather than having an auditor general, the finance and budget committees are, in the case of Senegal, assisted by the Court of Auditors (Cour des Comptes) in accordance with article 68 of the 2001 constitution. A similar framework is in place in Benin under article 37 of the Budget Act of 2001.

Institutional Capacity

A strong, dynamic, and effective parliament cannot exist without a parliamentary administration of equal quality. The organization of the parliamentary administration is a key component of a successful parliamentary institution. The internal organization and the provision of modern facilities and an improved information technology system are essential for building a strong parliamentary institution. Informed legislation and decision making rely on a parliament having strong policy analysis and research capacities. Recognizing that parliaments have different capacity levels, the API assesses the institutional capacity of parliament, which includes access to resources—human, material, and financial—to support MPs in the budget process. Figure 15.5 presents the scores that parliamentary participants gave their capacities.

Parliaments' ability to fairly and in a nonpartisan manner recruit competent staff and equip them with the needed resources was seen as a high-priority capacity indictor by all countries. Benin and Ghana, each with 7.9, have the highest score for this indicator. Ghana has a high score for human capacity but did not have adequate material and financial resources to fully execute legislative and oversight functions. Benin has the resources but not enough qualified staff members to support parliament's work. Kenya, which has the lowest capacity rating in this indicator (4.7), cannot determine its own budget, has an understaffed research department, and has basic logistics challenges.

Transparency and Integrity

Institutional integrity is fundamental to ensuring that the public believes and accepts parliament's decisions and actions. MPs and parliamentary staff members must be seen as above board in the performance of their responsibilities. Any negative perception of the legislature's integrity by the public will weaken and distort the authority and power balance between the executive and the legislature.

Figure 15.5 Average Weighted Capacity Ratio: Institutional Capacity of Parliament

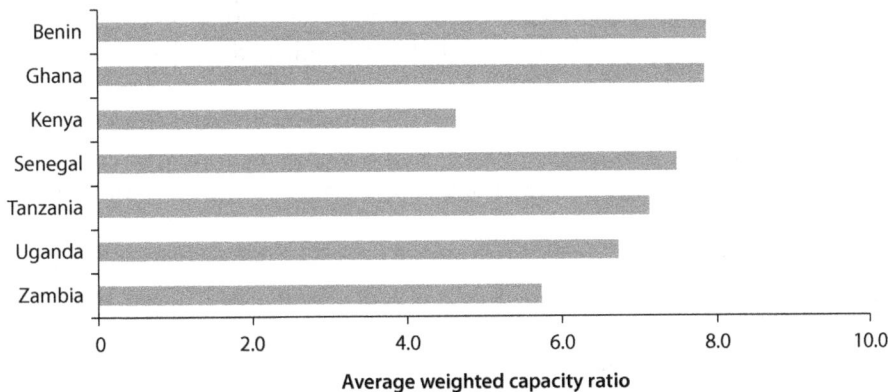

Source: African Parliamentary Index, Parliamentary Centre.

Benchmarking and Self-Assessment for Parliaments • http://dx.doi.org/10.1596/978-1-4648-0327-7

Figure 15.6 Weighted Capacity Ratio: Transparency and Integrity

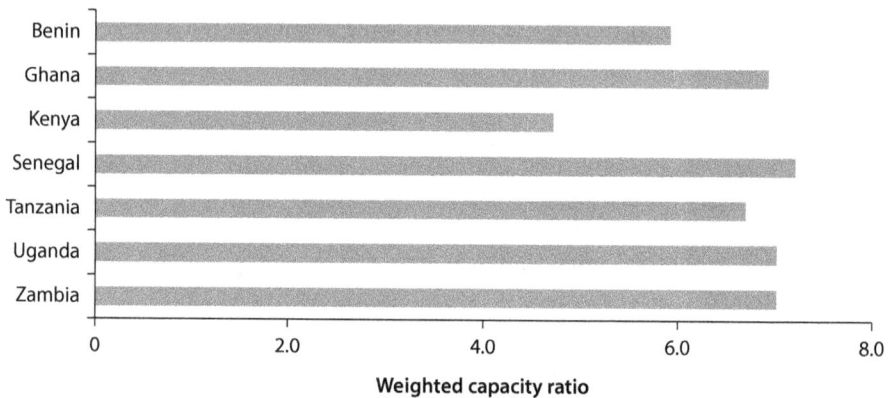

Source: African Parliamentary Index, Parliamentary Centre.

The legislature should therefore appeal to the conscience of MPs and parliamentary staff members to maintain high ethical standards in performing their duties. In this regard, this indicator assesses whether the legislature has a code of conduct and whether it is being enforced. It also examines whether the code is backed by legislation or a convention and whether it is published. Figure 15.6 indicates how parliaments themselves have put in place measures to encourage staff members and MPs to conduct themselves in honorable ways.

The relatively high-performing parliament in Senegal and the relatively low-performing parliament in Kenya equally rated this indictor as a priority to them. However, Senegal had taken more steps to provide a policy environment that ensures transparency and integrity of parliament. Information gathered from the assessment in Senegal revealed the existence of a code of conduct that is backed by legislation. The enforcement of legislation has seen the declaration of assets and private interest by some MPs, which means that there is still room for improvement to get members on board. In the case of Kenya, the legislature has no specific code of conduct, but some provisions in the standing orders guide the conduct of MPs. These provisions do not oblige members to declare their assets and business interest.

The perception of CSOs is that these two countries diverged from the self-assessments by the parliaments. Although CSOs scored the parliament in Kenya higher than parliamentarians scored themselves, CSOs scored the parliament in Senegal lower.

Overall Ranking

It is important to clarify that the API is a perception index based on the assumption that respondents are knowledgeable about their parliaments and will honestly score indicators on the basis of the descriptive guidelines provided for each subindicator. The API also assumes a similar geopolitical environment for all countries. As the previous analysis shows, the francophone countries clearly have a different institutional arrangement for oversight than the anglophone countries have. Thus, when ranking the overall capacity of the seven parliaments, one must be mindful of the geopolitical dynamics. Figure 15.7 shows the scores of the parliaments for each capacity area. Figure 15.8 indicates which countries might be creating an environment for best parliamentary practice.

Figure 15.7 Weighted Averages of Assessment Ratings per Capacity Area

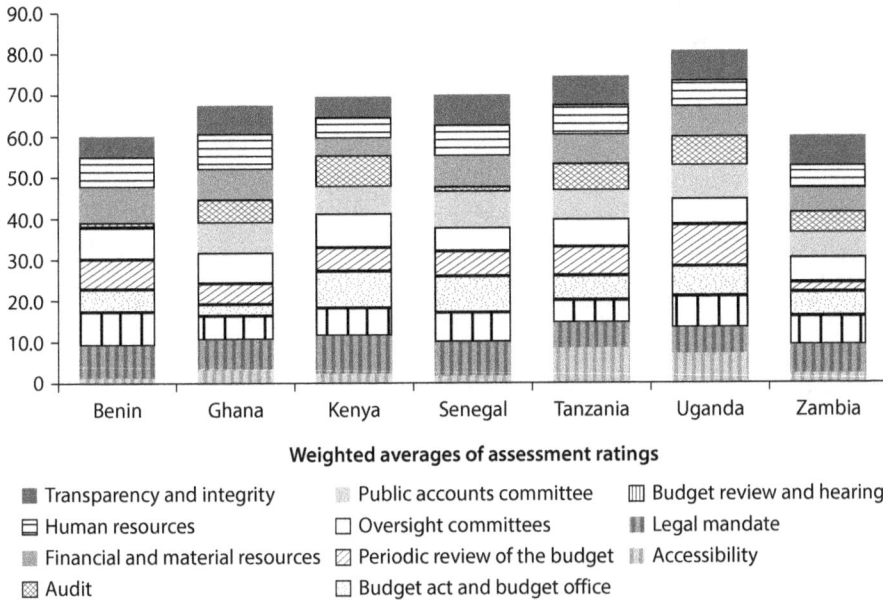

Weighted averages of assessment ratings

- ■ Transparency and integrity
- ▨ Public accounts committee
- ▥ Budget review and hearing
- ⊟ Human resources
- ☐ Oversight committees
- ▦ Legal mandate
- ▦ Financial and material resources
- ▨ Periodic review of the budget
- ▦ Accessibility
- ▨ Audit
- ☐ Budget act and budget office

Source: African Parliamentary Index, Parliamentary Centre.

Figure 15.8 African Parliamentary Index: Seven Country Rankings

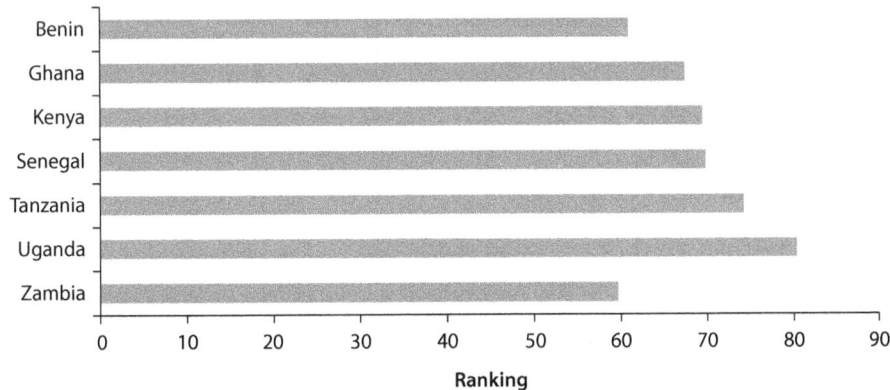

Source: African Parliamentary Index, Parliamentary Centre.

Benchmarking and Self-Assessment for Parliaments • http://dx.doi.org/10.1596/978-1-4648-0327-7

From the self-assessment, Uganda has the highest overall ranking (80.4), followed by Tanzania, Senegal, Kenya, Ghana, Benin, and Zambia. The reason for Uganda's high score is clear. Like Kenya, Uganda has enacted a budget act, which, since its introduction in 2001, has improved parliament's performance in the budget process. In line with the broad objectives articulated in article 155 of the Ugandan constitution, the act explicitly spells out the role of parliament in the budget process. The act facilitates increased flow of information relating to the national budget from government to parliament, which in turn aids periodic review of the budget. The act also established a budget office within the parliamentary service to collect, review, analyze, and report on budget-related information to all committees. Although the budget office may have some capacity challenges, it has nonetheless contributed greatly to parliament's improved capacity. The same can be said about Kenya, which passed the Fiscal Management Act and has a budget office.

In Tanzania, a highly decentralized system of planning and budgeting has contributed to improving citizens' participation in the budget process and access to parliament. Through the strategy of decentralization by devolution, which was introduced by a local government reform program in 1997, a system of local governance emerged that enables local government agencies to provide their mandated services to citizens in a transparent, accountable, accessible, equitable, and efficient manner. MPs are ex officio members of local councils, which also include representatives of wards. Thus, there is close contact between local representatives and MPs. As the intermediary between parliament and the citizens at the local level, the MPs disseminate, educate, and consult citizens on pertinent issues, including budget-related matters. In addition, regarding mechanisms to promote public understanding of the legislature's work, Tanzania has a parliamentary department on civic education, information, and international corporation whose duty is to ensure that the general public understands the legislature's work. In recent times, the parliament has enhanced the public understanding of its work through live television broadcasts.

Figure 15.9 shows the areas in which countries have scored high marks and which may be close to a best practice. These findings do not mean that these countries are doing perfectly well in terms of building capacity for their parliaments, but they have made some progress toward achieving the effective capacity required for a legislature to achieve its expected mandate.

According to figure 15.9, Benin had high scores for two indicators: (a) budget review and hearing and (b) financial and material resources. Ghana has skilled human resources and Senegal has capacity to enforce transparency and integrity among MPs and staff members. Uganda has capacity in three areas: (a) budget review and hearing, (b) periodic review of the budget, and (c) a particularly effective PAC. Kenya has capacity in four areas: (a) legal mandate, (b) budget act and budget office, (c) oversight committees, and (d) audit. Tanzania has the most enabling environment for citizens to access parliament.[1]

Figure 15.9 Areas of Capacity in Which Countries Are Close to Best Practice

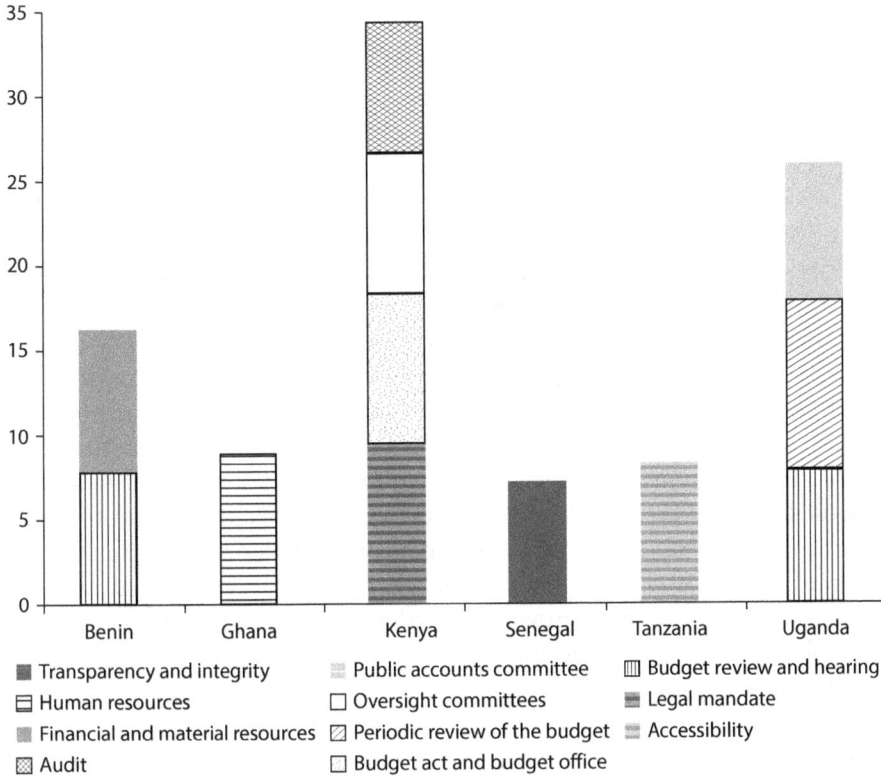

Legend:
- Transparency and integrity
- Human resources
- Financial and material resources
- Audit
- Public accounts committee
- Oversight committees
- Periodic review of the budget
- Budget act and budget office
- Budget review and hearing
- Legal mandate
- Accessibility

Source: African Parliamentary Index, Parliamentary Centre.

Conclusions and Recommendations

The findings of the API assessment reveal that parliaments with independent budget offices (Kenya and Uganda) received considerable support from these units, and the existence of these offices has in no small way led to the effectiveness of those parliaments with regard to budget oversight scrutiny. With enhanced capacities, parliamentarians can engage in informed debates and make cogent recommendations at committee sittings as well plenary sessions. Clearly, parliaments that lack such offices could benefit from their establishment.

A key function of parliament is representation of citizens, which involves collecting, aggregating, and expressing the concerns, opinions, and preferences of the country's citizens through the institution of parliament. The assessment results indicate that with the exception of Tanzania and Uganda, the partner-country parliaments fared poorly with respect to their accessibility to the public, particularly in relation to efforts being made by parliament to raise public awareness of its role and mandate. This finding, no doubt, calls for efforts to educate the public about the role of parliamentarians as well as inform the public about existing mechanisms available to citizens and the media to engage parliament more effectively.

The strength and effectiveness of parliament can also be measured by the extent to which parliament's operations are determined by itself rather than by the executive. The financial independence of parliament is crucial. The assessment results underlined the fact that all the partner parliaments, with the exception of Kenya, cannot determine and approve their own budgets. Reliance on the executive branch for funding and determination of salaries has the potential to severely compromise parliamentary autonomy. The leadership of the various parliaments is therefore encouraged to make strenuous efforts to engage the executive in a dialogue with a view to achieving this objective.

Parliament's mandate is typically derived from the country's constitution, which determines the formal rules of the political system and parliament's role and leverage therein. The assessment results point to a number of constitutional hurdles that confront parliaments. These parliaments have no power to amend proposed budgets submitted by the executive for parliamentary scrutiny and approval. The leverage thus exercised by parliament with respect to input made into the proposed budget is minuscule. In light of these challenges, constitutional reform needs to focus on improving the performance of parliaments.

A vibrant parliament is the cornerstone of democracy. Free and fair elections are an essential pillar, but elections must be accompanied by effective parliaments. Parliaments need to exert the constitutional powers they possess, and the all-too-common practice of viewing parliaments as a subbranch of the executive must be abandoned.

Note

1. Parliament self-assessments in most countries were disputed by the CSOs' validation assessment. Refer to country reports for the full API assessment and the conclusions reached.

CHAPTER 16

Other Benchmarking Experiences at the National Level

Lisa von Trapp

Introduction

A closer look at experiences at the national and state levels provides valuable insights into parliamentary assessment frameworks. Parliaments in Rwanda and Sierra Leone have used the Inter-Parliamentary Union (IPU) Self-Assessment Toolkit for Parliaments (IPU 2008) to bring fresh perspectives into their strategic planning exercises, and the Cambodian Senate used the IPU toolkit for its 10-year anniversary. A Pakistani think tank, the Pakistan Institute of Legislative Development and Transparency (PILDAT), used the IPU toolkit in partnership with Pakistan's parliament to carry out a nongovernmental organization assessment of parliament. The parliaments in Andorra and Ireland are currently using the toolkit to assess elements of their performance. An independent panel's assessment of South Africa's parliament included elements of the IPU toolkit, and the parliament's research unit also prepared a paper on measuring parliamentary performance that looked at examples from the work of the IPU, the Parliamentary Centre and the World Bank, and the Commonwealth Parliamentary Association (CPA). Finally, the toolkit has been tested with parliamentary administrations in Algeria and Sri Lanka (see chapter 10).

Several countries have also volunteered to "test" the CPA's (2006) Recommended Benchmarks for Democratic Legislatures. The first parliament to do so was the Australian Capital Territory Legislative Assembly, a regional parliament in Australia, in summer 2008 (see chapter 12). Since then, Canada has also undertaken a benchmarking self-assessment (see chapter 11). As part of the leadup to the Pacific Regional Benchmarks Meeting during the Forum Presiding Officers and Clerks Annual Meeting, the parliaments of Kiribati (see chapter 13), Nauru, Niue, and Tuvalu also undertook benchmarking exercises with the support of the United Nations Development Programme Pacific Centre. Staff members from South Australia's parliament prepared a benchmarking exercise, and staff members from the Federal Parliament of Australia used the IPU toolkit to contribute to discussions during a workshop on benchmarking of parliamentary performance for the Australian federal and state parliaments and the New Zealand parliament.

Lastly, the National Democratic Institute for International Affairs (NDI) piloted its questionnaire in Bosnia and Herzegovina in 2008 and, following revisions, administered the questionnaire in Colombia, Guatemala, Peru, and Serbia in 2009. Although it is not possible to fully review all of these case studies, the rest of this chapter presents snapshots from Cambodia, Colombia, Pakistan, and Rwanda.

Cambodia

The Cambodian Senate undertook a self-assessment exercise in 2009 (Sarith 2009). To oversee the exercise, the Cambodian Senate's Standing Committee established an ad hoc commission made up of the chairpersons of the nine specialized commissions (and representing all political parties), all department directors, and a number of experienced officials. Two working groups were formed: the first group was in charge of studying and answering questions from sections 1, 2, and 3 of the IPU toolkit, and the second group was in charge of sections 4, 5, and 6. Senators and senior officials actively participated, despite concerns about whether the evaluation should be public. The working groups' responses to the toolkit questions were submitted to the ad hoc commission for further review and improvement. The ad hoc commission then organized a three-day seminar attended by senators, parliamentary staff members from the Secretariat General, and international development partners. The seminar's purpose was to seek further recommendations. Lastly, the results were submitted to the Standing Committee for final approval.

The ad hoc commission and the working groups identified weak points to be addressed and developed a series of 15 reform recommendations for both the Senate and the Secretariat General. In particular, they recommended that the Senate organize visits and public consultations in the local commune (*sangkat*) to collect data and opinions on new law requirements and on the effect of existing laws. They also recommended that the Senate commissions devise clear and accurate work plans. Moreover, the secretary general was called on to increase the Senate's capacity by updating the strategic framework and plan of action and by continuing to seek assistance from other parliaments and development partners.

The Cambodian Senate saw these recommendations as a first step. In the medium and longer term, the ad hoc commission noted that it will have to continue its research on legal provisions and procedures stated in the constitution, internal regulations, Senate election laws, and the statutes of senators, as well as rules, duties, and competencies of the specialized commissions and the Secretariat General. The ad hoc commission also plans to study standards and parliamentary procedures regionally and globally.

Colombia

The NDI staff administered NDI's standards-based questionnaire in Colombia by using a guided one-on-one interview methodology.[1] There were a total of 39 participants: 24 legislators, 11 members of the parliamentary staff, and 4 civil society organization (CSO) representatives. The NDI staff attempted to select the most representative sample possible. Because Colombia's parliament is bicameral, NDI chose to test the questionnaire with members of the Chamber of Representatives (or lower house), because they have primary budget authority. NDI also helped to ensure that women participated in all three participant groups.

The interviews revealed the following preliminary conclusions:

- Perceptions of legislative power were relatively constant across the three target groups, but perceptions of legislative performance varied widely.
- Although all target groups found a gap between the power and practice of the legislature, CSO representatives perceived this gap as much wider than did legislators and legislative staff members.
- The gaps in perceptions of the three target groups also varied significantly depending on the legislative functions covered in the 25 two-part statements on power and practice.

The questionnaire revealed significant gaps between perceived power and practice in several areas, such as whether the legislature's committees have the power to summon materials and witnesses from the executive and whether they do so in practice (statements 7a and 7b on the questionnaire). There was also significant convergence in responses concerning the budget review process. For example, 75 percent of legislators and CSO representatives agreed that the legislature has the power to amend the national budget before approving it, but only 50 percent agreed that it has actually done so (NDI 2009, 16).

NDI plans to share the main findings from the questionnaire process with partners at the country level in the hope that the data will both contribute to dialogue among the target groups and support coordinated efforts to strengthen the functioning of the legislature. It is anticipated that the data will also be helpful to NDI field staff workers in their legislative strengthening efforts. For instance, because NDI staff members in Colombia work closely with party caucuses, they are particularly interested in seeing and sharing the data in that area.

Pakistan

In 2008, a prominent Pakistani think tank, PILDAT, undertook an evaluation of the National Assembly of Pakistan using the IPU self-assessment toolkit. PILDAT initiated the evaluation process by taking assembly leadership into confidence and requesting that parliamentarians participate in the evaluation group. The evaluation group included 14 parliamentarians from 5 political parties,

2 veteran parliamentary reporters, 3 senior academics, 2 senior journalists, 2 lawyers, 1 former military commander, and 2 PILDAT staff members (Karim Kundi 2009).

The evaluation group reviewed the six categories in the IPU self-assessment toolkit and noted an overall score for each section, as well as the weakest and strongest areas within each section. Based on these results, it then máde 11 recommendations:

- Make it possible for a person of average means to be elected to the parliament.
- Make the parliament's role in the budget process effective.
- Let parliament have a role in scrutinizing key appointments.
- Provide adequate and nonpartisan research service in the parliament.
- Institute systematic and transparent procedures for consulting citizen groups and experts while framing laws.
- Attract young people to the work of the National Assembly.
- Involve the public in the legislative process.
- Institute transparent and robust procedures and mechanisms to prevent conflicts of financial and other interest in the conduct of parliamentary business.
- Establish a system of adequate oversight over funding of candidates and parties during and after elections.
- Set up a system to monitor and review levels of public confidence in parliament.
- Have the Standing Committee on Foreign Affairs adopt a proactive role in formulating, shaping, and overseeing the foreign policy (PILDAT 2009).

Rwanda

Rwanda's parliament approached the IPU in 2008 to facilitate a self-assessment exercise as part of its review of its 2006–10 strategic plan (Power 2009). Because the parliament is bicameral, the self-assessment was done in two stages (with the Senate in December 2008 and the Chamber of Deputies in March 2009). The timing of the self-assessment fell around the middle of implementation of the strategic plan and followed on from the 2008 elections for the Chamber of Deputies, which led to a turnover of just over 50 percent of its members.

The self-assessment objectives were to identify (a) the parliament's strengths and weaknesses in the key strategic orientations, (b) ways to strengthen performance, and (c) ways to incorporate those elements in the parliament's strategic plan.[2] The process was slightly different for each chamber, but both went through the same five stages:

- An ad hoc committee was appointed to serve as the principal focus for the self-assessment exercise. Seven members were nominated but not all participated.

- The ad hoc committee had a half-day meeting to examine, amend, and adapt generic questions in the IPU toolkit to the Rwandan context.
- The amended questions were distributed to senators and members. In the case of the Senate, the ad hoc committee chair also convened a plenary session with a facilitator to explain the document to the 26 senators before asking them to complete the toolkit questionnaire. The Chamber of Deputies confined the use of the toolkit to the ad hoc committee members, thus limiting the representativeness of the sample.
- Results were collated and analyzed.
- Following deliberation, discussion, and recommendations by the ad hoc committee, the insights were applied to the operation and delivery of the strategic plan (Power 2009, 9–10).

The Rwandan parliament identified these four key issues and made recommendations related to each one:

- Recruiting, training, and retaining parliamentary staff members
- Ensuring that parliamentary procedures are understood and used by politicians (or "closing the gap between having and using powers")
- Changing the rules governing parliament in ways that will strengthen parliament (for example, better scrutiny by committees of legislation's implementation)
- Monitoring and implementing the strategic plan (for example, regular progress reviews or reports by the parliamentary bureau in conjunction with the secretary general) (Power 2009, 2)

Notes

1. Some legislative staff members and CSO representatives were permitted to fill out the questionnaire individually if the field staff believed that they would be more likely to present their true views while on their own.

2. The strategic plan has six strategic orientations: (a) improving the legislative process, (b) strengthening oversight, (c) effectively supervising the fundamental principles of the constitution, (d) improving communication, (e) promoting parliamentary diplomacy and dialogue, and (f) developing parliament's administrative capacity (Power 2009, 3).

References

CPA (Commonwealth Parliamentary Association). 2006. "Recommended Benchmarks for Democratic Legislatures." CPA, London. http://wbi.worldbank.org/wbi/Data/wbi/wbicms/files/drupal-acquia/wbi/Recommended%20Benchmarks%20for%20Democratic%20Legislatures.pdf.

IPU (Inter-Parliamentary Union). 2008. "Evaluating Parliament: A Self-Assessment Toolkit for Parliaments." IPU, Geneva. http://www.ipu.org/pdf/publications/self-e.pdf.

Karim Kundi, Faisal. 2009. "Evaluation of the National Assembly of Pakistan." Presented at the Inter-Parliamentary Union–Association of Secretaries General of Parliaments Conference on Evaluating Parliament, Objectives, Methods, and Results, Geneva, October 22.

NDI (National Democratic Institute for International Affairs). 2009. "Final Report: Questionnaire on International Standards for the Functioning of Democratic Legislatures." Grant report prepared for the World Bank, Washington, DC.

PILDAT (Pakistan Institute of Legislative Development and Transparency). 2009. "State of Democracy in Pakistan: Evaluation of Parliament, 2008–2009." PILDAT, Islamabad. http://www.agora-parl.org/sites/default/files/evaluationofparliament2008-2009.pdf.

Power, Greg. 2009. "The Rwandan Parliament's Self-Assessment Exercise: Insights and Issues." Inter-Parliamentary Union, Geneva.

Sarith, Oum. 2009. "Evaluating Parliament: Objective, Methods, Results, and Impact—Senate of the Kingdom of Cambodia." Presented at the Inter-Parliamentary Union–Association of Secretaries General of Parliaments Conference on Evaluating Parliament, Objectives, Methods, and Results, Geneva, October 22.

green
press
INITIATIVE

www.ingramcontent.com/pod-product-compliance
Lightning Source LLC
Chambersburg PA
CBHW080412270326
41929CB00018B/3002